AF594937

Toxicology and Biocompatibility of Nanomaterials

Toxicology and Biocompatibility of Nanomaterials

Editors

Carsten Weiss
Silvia Diabaté

MDPI • Basel • Beijing • Wuhan • Barcelona • Belgrade • Manchester • Tokyo • Cluj • Tianjin

Editors

Carsten Weiss
Institute of Biological and Chemical Systems – Biological Information Processing (IBCS-BIP)
Karlsruhe Institute of Technology (KIT)
Eggenstein-Leopoldshafen
Germany

Silvia Diabaté
Institute of Biological and Chemical Systems – Biological Information Processing (IBCS-BIP)
Karlsruhe Institute of Technology (KIT)
Eggenstein-Leopoldshafen
Germany

Editorial Office
MDPI
St. Alban-Anlage 66
4052 Basel, Switzerland

This is a reprint of articles from the Special Issue published online in the open access journal *Nanomaterials* (ISSN 2079-4991) (available at: www.mdpi.com/journal/nanomaterials/special_issues/toxic_biocompt).

For citation purposes, cite each article independently as indicated on the article page online and as indicated below:

LastName, A.A.; LastName, B.B.; LastName, C.C. Article Title. *Journal Name* **Year**, *Volume Number*, Page Range.

ISBN 978-3-0365-2739-0 (Hbk)
ISBN 978-3-0365-2738-3 (PDF)

Contents

About the Editors

Carsten Weiss

Carsten Weiss is a research group leader at the Institute of Chemical and Biological Systems at the Karlsruhe Institute of Technology (KIT). Upon completing his PhD in Biology at the University of Karlsruhe, he gained practical experience in the field of Molecular Toxicology, including the completion of several post-doctoral fellowships in Germany and the USA. His main research addresses the role of signaling in response to environmental particulate matter and genotoxins. Dr. Weiss was and still is involved in numerous national and European projects concerned with safety assessment and the molecular action of (nano)materials and genotoxins.

Silvia Diabaté

Dr. Silvia Diabaté is a biologist by training and has extensive experience in the toxicology of combustion-derived particles and nanoparticles, with special emphasis on inflammatory and anti-oxidative processes in lung cells. Together with her colleagues at KIT, she developed in vitro assays for investigating the potential toxicity of aerosols in cell cultures after exposure at the air–liquid interface.

Preface to "Toxicology and Biocompatibility of Nanomaterials"

With great pleasure, we served as guest editors for this Special Issue entitled "Toxicology and Biocompatibility of Nanomaterials". Although the benefits of nanomaterials are obvious, concern has also been raised about such materials being inhaled, ingested, applied to the skin, or even released into the environment, potentially inducing adverse effects. In this book, some timely original research articles as well as reviews address various aspects of this interesting topic. Once more, we thank all of the authors for their valuable contributions, and the many constructive reviewers and the editorial team who helped to bring this Special Issue to fruition. We hope that a broad readership will enjoy this book.

Carsten Weiss, Silvia Diabaté
Editors

Editorial

Toxicology and Biocompatibility of Nanomaterials

Carsten Weiss * and Silvia Diabaté

Institute of Biological and Chemical Systems-Biological Information Processing, Karlsruhe Institute of Technology (KIT), Hermann-von-Helmholtz-Platz 1, 76344 Eggenstein-Leopoldshafen, Germany; silvia.diabate@kit.edu
* Correspondence: carsten.weiss@kit.edu

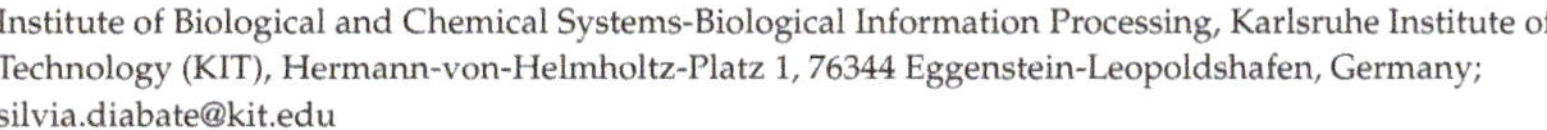

It is our great pleasure to introduce this Special Issue entitled "Toxicology and Biocompatibility of Nanomaterials". As the understanding of materials at the nanoscale and the ability to control their structure improves, a wide range of nanomaterials (NMs) with novel characteristics and applications are being fabricated for electronics, engineering, and, more recently, biomedical research applications. Although the technological and economic benefits of NMs are obvious, concern has also been raised that the very same properties, which enable a variety of novel applications, might have adverse effects if such a material is inhaled, ingested, applied to the skin or even released into the environment [1]. These concerns have led to an increasing discussion worldwide about possible regulatory policies for NMs. Therefore, there is a clear need to establish convincing scientific knowledge to assess the impact of NMs on human health and the ecosystem. These questions can only be tackled by collaborative research at the interface of engineering, physics, chemistry, toxicology and biology, as outlined in our first Special Issue on the topic one decade ago [2]. Meanwhile, the field of nanotoxicology has come of age, and is an established discipline in toxicology.

Citation: Weiss, C.; Diabaté, S. Toxicology and Biocompatibility of Nanomaterials. *Nanomaterials* **2021**, *11*, 3110. https://doi.org/10.3390/nano11113110

Received: 11 November 2021
Accepted: 17 November 2021
Published: 18 November 2021

This Special Issue comprises six research articles and two timely reviews and covers research in the field of nanotoxicology, with a particular interest in molecular mechanism of action as well as the safe-by-design concept, i.e., the synthesis of biocompatible nanomaterials. Additionally, the impact of the biomolecular corona, which is the interaction of biomolecules with the NM surface, on toxicity and biocompatibility, is addressed.

The first study addresses the effects of ZnO NMs on blood glucose levels in healthy and diabetic rats and discusses potential clinical applications [3]. The following articles are mainly in vitro studies with a focus on mechanisms of NM toxicity. Macrophages are important targets of NMs; therefore, the understanding of molecular initiating events is of high relevance. A series of TiO_2 NMs with different characteristics were studied in rat alveolar macrophages and activation of the NLRP3 inflammasome is presented, highlighting the importance of case-to-case studies for proper hazard assessment [4]. In murine macrophages, the poorly understood mixture effects of co-pollutants and NMs has been investigated [5]. The authors demonstrate synergistic activities of silica NMs and genotoxic agents, thus reinforcing the notion that there is an urgent need to pay more attention to mixture effects in the future. Apart from the innate immune system, the gastrointestinal tract is an important target tissue. Therefore, the toxicity of silica NMs is explored in gastrointestinal cells [6], specifically in the presence or absence of serum, because the biomolecular corona has previously been shown to critically determine detrimental effects in other cell types [7]. Although in the presence of serum even pro-proliferative effects have been shown in gastric cells [8], in the absence of serum, silica NMs potently induce cell death in colon carcinoma cells; however, this is independent of the key regulators p53 and BAX, suggesting the potential for their further development as anticancer nanodrugs [6]. Although the field of nanotoxicology has matured and solved most of the initial technical problems and challenges, there are still issues which hamper proper hazard assessment. These include, but are not limited to, the reproducible synthesis of NMs with clearly characterized physico-chemical properties [9], as well as the establishment of more physiologically relevant test systems [10] and standard operating procedures for toxicity testing to provide comparable results across laboratories. Thus, fifteen European laboratories performed an inter-laboratory comparison to assess the toxicity of polystyrene NMs with the widely used MTS assay and provide guidance on how to improve reliable testing [11]. Finally, the interactions of CeO_2 and TiO_2 NMs and algae were addressed in an ecotoxicity study, and the importance of the adherence of NMs to the test organism was identified as an important parameter to predict toxicity [12].

Two reviews conclude this Special Issue. The first summarizes our knowledge of airborne NMs and their potential adverse effects on the nervous system with a specific focus on neurodegenerative diseases [13]. The final contribution provides a fresh outlook on positive aspects of NM actions in biological systems, i.e., their use in medicine as tools to diagnose and treat cancer [14].

In conclusion, we would like to thank all the authors for their interesting and excellent contributions, and the many constructive reviewers and the editorial team who helped to bring this Special Issue to fruition. We hope that a broad readership will enjoy this Special Issue.

Author Contributions: C.W. prepared the original draft, which was reviewed and edited by S.D. All authors have read and agreed to the published version of the manuscript.

Funding: This article received no external funding.

Data Availability Statement: Not applicable.

Acknowledgments: We are grateful to Mirabelle Wang from *Nanomaterials* for her continuous support. We are grateful to all the authors for submitting their studies to the present Special Issue and for its successful completion. We deeply acknowledge the *Nanomaterials* reviewers for enhancing the quality and impact of all submitted papers.

Conflicts of Interest: The authors declare no conflict of interest.

References

1. Gebel, T.; Foth, H.; Damm, G.; Freyberger, A.; Kramer, P.J.; Lilienblum, W.; Rohl, C.; Schupp, T.; Weiss, C.; Wollin, K.M.; et al. Manufactured nanomaterials: Categorization and approaches to hazard assessment. *Arch. Toxicol.* **2014**, *88*, 2191–2211. [CrossRef] [PubMed]
2. Weiss, C.; Diabate, S. A special issue on nanotoxicology. *Arch. Toxicol.* **2011**, *85*, 705–706. [CrossRef] [PubMed]
3. Virgen-Ortiz, A.; Apolinar-Iribe, A.; Diaz-Reval, I.; Parra-Delgado, H.; Limon-Miranda, S.; Sanchez-Pastor, E.A.; Castro-Sanchez, L.; Jesus Castillo, S.; Dagnino-Acosta, A.; Bonales-Alatorre, E.; et al. Zinc Oxide Nanoparticles Induce an Adverse Effect on Blood Glucose Levels Depending on the Dose and Route of Administration in Healthy and Diabetic Rats. *Nanomaterials* **2020**, *10*, 2005. [CrossRef] [PubMed]
4. Kolling, J.; Tigges, J.; Hellack, B.; Albrecht, C.; Schins, R.P.F. Evaluation of the NLRP3 Inflammasome Activating Effects of a Large Panel of TiO_2 Nanomaterials in Macrophages. *Nanomaterials* **2020**, *10*, 1876. [CrossRef] [PubMed]
5. Dussert, F.; Arthaud, P.A.; Arnal, M.E.; Dalzon, B.; Torres, A.; Douki, T.; Herlin, N.; Rabilloud, T.; Carriere, M. Toxicity to RAW264.7 Macrophages of Silica Nanoparticles and the E551 Food Additive, in Combination with Genotoxic Agents. *Nanomaterials* **2020**, *10*, 1418. [CrossRef] [PubMed]
6. Fritsch-Decker, S.; An, Z.; Yan, J.; Hansjosten, I.; Al-Rawi, M.; Peravali, R.; Diabate, S.; Weiss, C. Silica Nanoparticles Provoke Cell Death Independent of p53 and BAX in Human Colon Cancer Cells. *Nanomaterials* **2019**, *9*, 1172. [CrossRef] [PubMed]
7. Leibe, R.; Hsiao, I.L.; Fritsch-Decker, S.; Kielmeier, U.; Wagbo, A.M.; Voss, B.; Schmidt, A.; Hessman, S.D.; Duschl, A.; Oostingh, G.J.; et al. The protein corona suppresses the cytotoxic and pro-inflammatory response in lung epithelial cells and macrophages upon exposure to nanosilica. *Arch. Toxicol.* **2019**, *93*, 871–885. [CrossRef] [PubMed]
8. Wittig, A.; Gehrke, H.; Del Favero, G.; Fritz, E.M.; Al-Rawi, M.; Diabate, S.; Weiss, C.; Sami, H.; Ogris, M.; Marko, D. Amorphous Silica Particles Relevant in Food Industry Influence Cellular Growth and Associated Signaling Pathways in Human Gastric Carcinoma Cells. *Nanomaterials* **2017**, *7*, 18. [CrossRef] [PubMed]
9. Mulhopt, S.; Diabate, S.; Dilger, M.; Adelhelm, C.; Anderlohr, C.; Bergfeldt, T.; Gomez de la Torre, J.; Jiang, Y.; Valsami-Jones, E.; Langevin, D.; et al. Characterization of Nanoparticle Batch-To-Batch Variability. *Nanomaterials* **2018**, *8*, 311. [CrossRef] [PubMed]
10. Diabate, S.; Armand, L.; Murugadoss, S.; Dilger, M.; Fritsch-Decker, S.; Schlager, C.; Beal, D.; Arnal, M.E.; Biola-Clier, M.; Ambrose, S.; et al. Air-Liquid Interface Exposure of Lung Epithelial Cells to Low Doses of Nanoparticles to Assess Pulmonary Adverse Effects. *Nanomaterials* **2020**, *11*, 65. [CrossRef] [PubMed]

11. Nelissen, I.; Haase, A.; Anguissola, S.; Rocks, L.; Jacobs, A.; Willems, H.; Riebeling, C.; Luch, A.; Piret, J.P.; Toussaint, O.; et al. Improving Quality in Nanoparticle-Induced Cytotoxicity Testing by a Tiered Inter-Laboratory Comparison Study. *Nanomaterials* **2020**, *10*, 1430. [CrossRef] [PubMed]
12. Hund-Rinke, K.; Sinram, T.; Schlich, K.; Nickel, C.; Dickehut, H.P.; Schmidt, M.; Kuhnel, D. Attachment Efficiency of Nanomaterials to Algae as an Important Criterion for Ecotoxicity and Grouping. *Nanomaterials* **2020**, *10*, 1021. [CrossRef] [PubMed]
13. Von Mikecz, A.; Schikowski, T. Effects of Airborne Nanoparticles on the Nervous System: Amyloid Protein Aggregation, Neurodegeneration and Neurodegenerative Diseases. *Nanomaterials* **2020**, *10*, 1349. [CrossRef] [PubMed]
14. Siemer, S.; Wunsch, D.; Khamis, A.; Lu, Q.; Scherberich, A.; Filippi, M.; Krafft, M.P.; Hagemann, J.; Weiss, C.; Ding, G.B.; et al. Nano Meets Micro-Translational Nanotechnology in Medicine: Nano-Based Applications for Early Tumor Detection and Therapy. *Nanomaterials* **2020**, *10*, 383. [CrossRef] [PubMed]

Communication

Zinc Oxide Nanoparticles Induce an Adverse Effect on Blood Glucose Levels Depending on the Dose and Route of Administration in Healthy and Diabetic Rats

Adolfo Virgen-Ortiz [1,*], Alejandro Apolinar-Iribe [2], Irene Díaz-Reval [1], Hortensia Parra-Delgado [3], Saraí Limón-Miranda [4], Enrique Alejandro Sánchez-Pastor [1], Luis Castro-Sánchez [5], Santos Jesús Castillo [6], Adan Dagnino-Acosta [5], Edgar Bonales-Alatorre [1] and Alejandrina Rodríguez-Hernández [7]

1 Centro Universitario de Investigaciones Biomédicas, Universidad de Colima, Colima C.P. 28045, Mexico; idiazre@ucol.mx (I.D.-R.); espastor@ucol.mx (E.A.S.-P.); ebonales0@ucol.mx (E.B.-A.)
2 Departamento de Física, Universidad de Sonora, A.P. 1626, Hermosillo, Sonora C.P. 83000, Mexico; apolinar@ciencias.uson.mx
3 Facultad de Ciencias Químicas, Universidad de Colima, Coquimatlán, Colima C.P. 28400, Mexico; hparra@ucol.mx
4 Departamento de Ciencias Químico Biológicas y Agropecuarias, URS, Universidad de Sonora, Navojoa, Sonora C.P. 85880, Mexico; sarai.limon@unison.mx
5 Centro Universitario de Investigaciones Biomédicas, CONACYT-Universidad de Colima, Universidad de Colima, Colima C.P. 28045, Mexico; luis_castro@ucol.mx (L.C.-S.); dagninoa@ucol.mx (A.D.-A.)
6 Departamento de Investigación en Física, A.P. 5-088, Hermosillo, Sonora C.P. 83000, Mexico; santos.castillo@unison.mx
7 Facultad de Medicina, Universidad de Colima, Colima C.P. 28040, Mexico; arodrig@ucol.mx
* Correspondence: avirgen@ucol.mx

Received: 12 September 2020; Accepted: 4 October 2020; Published: 12 October 2020

Abstract: Different studies in experimental diabetes models suggest that zinc oxide nanoparticles (ZnONPs) are useful as antidiabetic agents. However, this evidence was performed and measured in long-term treatments and with repeated doses of ZnONPs. This work aimed to evaluate the ZnONPs acute effects on glycemia during the next six h after an oral or intraperitoneal administration of the treatment in healthy and diabetic rats. In this study, the streptozotocin-nicotinamide intraperitoneal administration in male Wistar rats were used as a diabetes model. 10 mg/kg ZnONPs did not modify the baseline glucose in any group. Nevertheless, the ZnONPs short-term administration (100 mg/kg) induced a hyperglycemic response in a dose and route-dependent administration in healthy (130 ± 2 and 165 ± 10 mg/dL with oral and intraperitoneal, respectively) and diabetic rats (155 ± 2 and 240 ± 20 mg/dL with oral, and intraperitoneal, respectively). The diabetic rats were 1.5 fold more sensitive to ZnONPs effect by the intraperitoneal route. In conclusion, this study provides new information about the acute response of ZnONPs on fasting glycemia in diabetic and healthy rat models; these data are essential for possible future clinical approaches.

Keywords: zinc oxide nanoparticles; diabetes; hyperglycemic response; zinc; nanomedicine; nanoparticle toxicology

1. Introduction

Worldwide, a decade ago, more than 30,000 t of ZnONPs were produced annually [1], surely today it has increased significantly due to its wide industrial application in cosmetics, sunscreens, coatings, paints and antimicrobials. ZnONPs have shown catalytic, electrical, photochemical, anticorrosive, photovoltaic, antifungal, antibacterial, and antiviral activity [2]. In the biomedical field, ZnONPs have been used for development of biosensors for a wide variety of molecules of interest, to improve diagnosis through imaging, controlled drug release, gene delivery and as therapeutic agents [3–6]. There is promising scientific evidence for the treatment of diseases with a high worldwide prevalence where several studies evaluated the anticancer, antidiabetic and antimicrobial activity of ZnONPs [7–9]. The whole potential of the application of these nanoparticles for the benefit of humans, demands studies and a detailed understanding of all their possible toxic or adverse effects on human health and the environment. In the literature, some toxic effects of ZnONPs have been described, these vary according to factors, such as their physicochemical characteristics, the concentrations, doses, exposure time and the route of administration used in the experiments. In general, it is suggested that the toxic effects produced by ZnONPs in different tissues or cell lines are mediated by increased oxidative stress and inflammation [7,10].

On the other hand, in the balance of benefits versus toxicology of metal nanoparticles for the treatment of mellitus diabetes, different studies have described that metal nanoparticles (silver, gold, cooper, selenium, magnesium, cerium oxide, titanium dioxide [11–18] and zinc oxide possess antihyperglycemic activity in diabetic rats after daily treatment for different periods [8,19–23].

In particular, although ZnONPs have been reported to have antihyperglycemic activity, studies in this regard are scarce. For this reason, more detailed research is required to determine their importance as therapeutic agents in chronic treatments. A recent study, demonstrated that an oral administration of 1–10 mg/kg during 4 weeks reduced hyperglycemia in type 1 diabetes (D1) and type 2 diabetes (D2), but the insulin level was not affected in D1. In contrast, insulin levels only increased at a dose of 10 mg/kg in D2, explaining the improved glucose tolerance in this model [19]. A similar effect was observed after a seven-weeks treatment [22]. However, other related studies using the identical administration route during four or eight weeks showed an increase in insulin levels associated with the antihyperglycemic effects in D1 [8,23]. The critical evidence supporting that oral nanoparticles administration in a dose range of 1–10 mg/kg/day for several weeks has antidiabetic activity; nonetheless, the immediate effect ZnONPs post-administration on basal glycemia has not been studied and its evaluation is essential to detect a possible risk in the diabetic patient since both an increase or a drastic decrease in glucose levels compromises their health., e.g., the acute hypoglycemia and hyperglycemia have been reported to induce atherothrombotic effects in non-diabetic and diabetic individuals, and these alterations have been associated with an increase in morbidity and mortality caused by cardiovascular failure [24–26]. Furthermore, an acute drastic imbalance in blood glucose levels in a diabetic patient can induce a potentially fatal diabetic coma [27].

The lack of this information limits the integral control of alterations suffered by diabetic patient. The goal of this research was to evaluate the acute effects on glycemia of oral and intraperitoneal administration of ZnONPs in healthy and diabetic rats.

2. Materials and Methods

2.1. Material

Zinc oxide nanoparticles dispersion (Cat. No. 721077, density 1.7 g/mL), nicotinamide (Cat. No. N3376, purity ≥ 98% HPLC), and streptozotocin (Cat. S0130, purity ≥ 98% HPLC) were obtained from Sigma-Aldrich Co. (St Louis, MO, USA); a sterile saline solution of sodium chloride (0.9%) was acquired from PISA Pharmaceutical Co. (Jalisco, Mexico).

It has been reported that ZnONPs in aqueous media is unstable depending on the concentration, pH and ionic strength of the medium [28]. However, it has also been described that sonication produces a stable suspension useful for biological assays [29]. Therefore, in our study for in vivo tests, a suspension of ZnONPs was freshly prepared at a concentration of 10 mg/mL as follows: The ZnONPs were deposited in a sterile saline solution and subsequently sonicated for 10 min (50% pulse amplitude with resting times of 30 seconds between pulses, 130 Watts, 20 KHz Ultrasonic Processor (Cole-Palmer Instruments, Vernon Hills, IL, USA).

2.2. ZnONPs Characterization

The shape and size of the nanoparticles were determined using scanning transmission electron microscopy (STEM, JEOL, JSM-7800F, Pleasanton, CA, USA) in an aliquot of ZnONPs suspension. The hydrodynamic diameter was measured in a previously sonicated suspension of ZnONPs (dissolved in 0.9% NaCl), the measurement based on dynamic light scattering (DLS) was performed using a ZetaPlus size analyzer (Brookhaven Instruments Co., Holtsville, NY, USA).

2.3. Animals

For this study, intact three-month-old male Wistar rats (n = 96) were used, this particular age was selected to reduce the streptozotocin sensitivity, since this drug induces experimental diabetes with known higher sensitivity in very young rats [30].

The rats were maintained in individual cages with water and food *ad libitum* (Rodent Laboratory Chow 5001, PMI Nutrition International LLC). They were kept in a room with light-dark cycles (12 h/12 h) and room temperature control (25 °C). During a week prior to the start of the experiment, all rats were manipulated for their adaptation and to eliminate manipulation stress at the time of performing glucose measurements in vivo. All studies were conducted in accordance with the Guide for the Care and Use of Laboratory Animals published by the US National Institute of Health (NIH) and approved by the Bioethics Committee of the University of Colima (Approval number 2018-15).

2.4. Experimental Design and Diabetes Induction

The rats were divided into 2 groups: Diabetic rats (n = 48) and non-diabetic rats (n = 48). Experimental diabetes in rats was induced by an intraperitoneal sequential treatment with streptozotocin and nicotinamide. First, streptozotocin was dissolved in citrate buffer pH = 4.5 and then administrated (65 mg/kg body weight). After fifteen min, nicotinamide dissolved in 0.9% saline solution was injected (230 mg/kg body weight). This model induces partial cytotoxicity on pancreatic β-cells producing moderate hyperglycemia without body weight loss or drastic decreases of plasma insulin levels [31]. After seven days, the glycemia was measured in blood samples collected from rat tail using an Accu-chek® Performa blood glucose system (Roche Diagnostics, Mannnheim, Germany); rats with fasting glucose of 126 mg/dL were included in the diabetes group (World Health Organization).

2.5. Evaluation of Zinc Oxide Nanoparticles on Fasting Glycemia Values

All rats were fasted for 8 h (07:00 am–03:00 pm) before evaluation. Both groups, diabetic and non-diabetic rats, were subdivided (n = 8 by subgroup) for the test of two doses of ZnONPs, 10 and 100 mg/kg body weight by two administration routes, oral or intraperitoneal. Before each administration, the ZnONPs dispersion was previously vortexed for 30 seconds to maintain its homogeneity.

Glycemia was evaluated at time 0 and 15, 30, 60, 90, 120, 240, and 360 min ZnONPs post-administration using an Accu-chek® Performa blood glucose monitor (Roche Diagnostics, Mannnheim, Germany). The blood sample was obtained from the distal part of pre-cleaned rat tail using an alcohol swab; immediately after, a small cut was made with scissors and the blood obtained is deposited on the test strip and placed on the digital glucometer. The clot was removed for future fresh blood collection to perform the glucose measurement. This procedure is repeated with each rat.

2.6. Statistics

All data is expressed as mean ± standard error. Experimental results were analyzed using a one-way ANOVA with post hoc test (Bonferroni) for statistical differences among groups. Differences with $p < 0.05$ were considered significant.

3. Results and Discussion

The images obtained by STEM demonstrated that the ZnONPs have a spherical shape, and the size analysis performed with ImageJ software determined that they have an average diameter of 17 ± 3.6 nm (Figure 1). DLS analysis revealed that ZnONPs dissolved in saline solution have an average hydrodynamic diameter of 1455 nm and polydispersity index of 0.48.

Figure 1. STEM imagen of ZnONPs shows a spherical shape. The scale bar represent a 100 nm length.

ZnONPs dispersion intraperitoneally administered with a 100 mg/kg single dose generated a significant increase in glycemia, compared with the control group treated with vehicle ($p < 0.05$), reaching a maximum peak 30 min after the administration in healthy rats (Figure 2B) and 60 min after in diabetic rats (Figure 3B). The increased levels of blood glucose returned basal levels 6 h post-administration and reached higher levels in diabetic rats when compared with the healthy control group ($p < 0.05$). In contrast, the low dosage tested in this study of 10 mg/kg ZnONPs intraperitoneally administered with same conditions described above generated undistinguishable effects at least in the time range monitored of 6 h ($p > 0.05$) in healthy and diabetic rats.

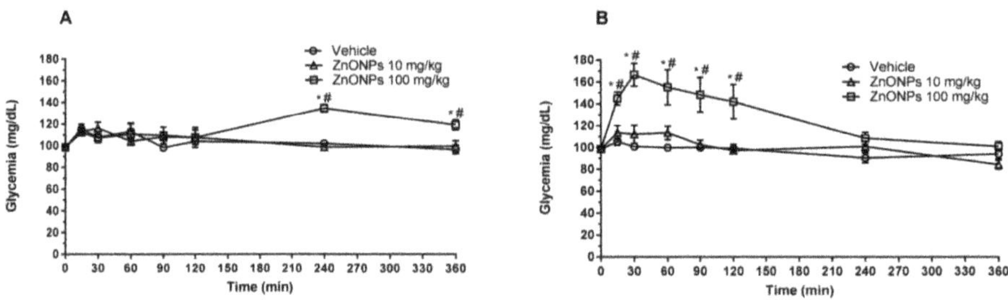

Figure 2. Short-term effects on glycemia of oral or intraperitoneal administration of ZnONPs in fasted healthy rats. (**A**) Oral administration (p.o). (**B**) Intraperitoneal route (i.p). Vehicle (Sterile 0.9% sodium chloride solution). * significant in comparison with vehicle ($p < 0.05$, n = 8 by group), # significant difference in comparison with group treated (10 mg/kg).

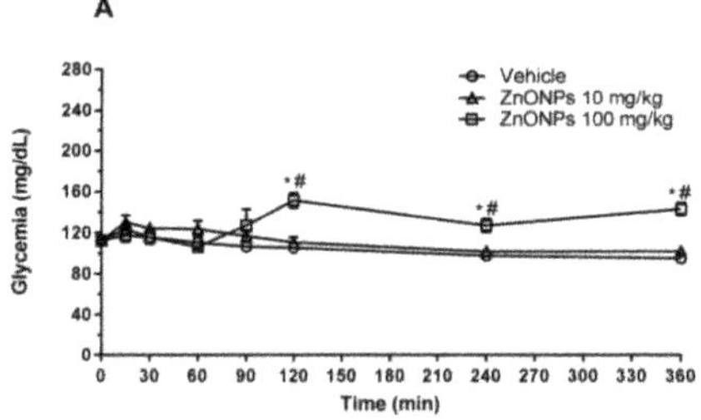

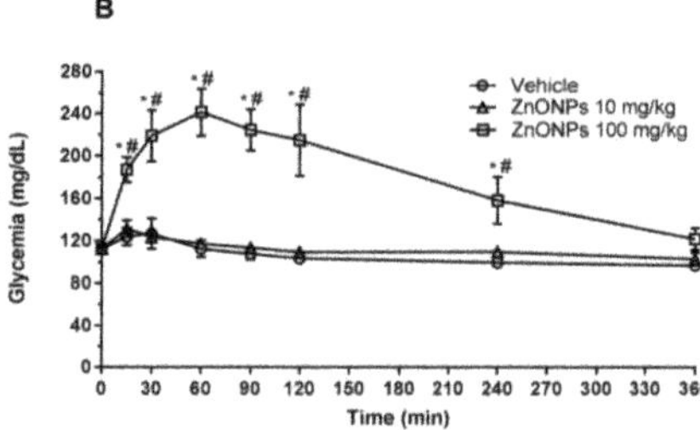

Figure 3. Short-term effects on glycemia of oral or intraperitoneal administration of ZnONPs in fasted diabetic rats. (**A**) Oral administration (p.o). (**B**) Intraperitoneal route (i.p). Vehicle (Sterile 0.9% sodium chloride solution). * significant in comparison to vehicle ($p < 0.05$, n = 8 by group), # significant in comparison with group treated with 10 mg/kg.

Oral administration of ZnONPs (100 mg/kg) significantly increased the glucose levels with a lag of 2 h after the administration. This increase was sustained for additional 4 h, in both healthy (Figure 2A) and diabetic rats (Figure 3A).

ZnONPs administered intraperitoneally (100 mg/kg) to diabetic rats showed an hyperglycemic induction significantly higher than oral supplementacion using the same dose. However, six hours post-administration, the i.p group showed a normal glycemia with no significant differences in comparison with the oral administration group (Figures 2B and 3B).

As mentioned in the background, it is widely described that ZnONPs have antidiabetic activity when administered for long periods of time. However, their immediate post-administration effects were not studied before. We tested whether they had a hypoglycemic effect with two different doses or affects related to the route of administration. To our surprise, the results showed that ZnONPs do not produce hypoglycemic effects in the short-term, and on the contrary, induce a hyperglycemic response depending on the dose, route of administration and health status (diabetes). To our knowledge, this is the first report demonstrating the ability of ZnONPs to generate short-term hyperglycemic response through a currently unknown mechanism.

The anti-diabetic activity of ZnONPs in the long term is proposed to be carried out as a result of the stimulation of several mechanisms, among them are the suggestion of an increase of serum insulin levels, glucokinase activity, and increased of insulin, insulin receptor A, GLUT-2 (Glucose transporter 2), and glucokinase mRNA (messenger Ribonucleic Acid) expression [8], reduction in oxidative stress [19,22], less damage to the pancreatic structure [32,33] and microRNA-103 and microRNA-143 decreased expression [23]. In vitro experiments revealed that ZnONPs attenuate the hyperglycemia through a mechanism that involves α-amylase inhibition and α-glucosidase activity [34]. Moreover, in vitro experiments showed that ZnONPs induce GLUT-4 (Glucose transporter 4) translocation and increase β-cell proliferation [35].

In contrast, the lack of knowledge about the mechanism involved in the short-term hyperglycemic response induced by ZnONPs generates new research questions for future work. It is widely known that the liver is the main organ generator of free glucose and is an essential target in antidiabetic therapies [36]. The hyperglycemic effect reported in the present work could be the result of a direct action of a high concentration of zinc ions on the hepatic metabolism; in hepatocytes, zinc at high levels induces an increase of glucose production through glycogenolysis [37]. The zinc supplementation in rats produces a hyperglycemic response, an increase of glucagon, a decrease of insulin, depletion of hepatic glycogen, and hyperglycemia attenuation when the adrenal glands were previously removed [38]. Despite these studies, the action mechanism evidence of a hyperglycemic response by zinc supplementation in the short-term is insufficient, and our results with ZnONPs increase the interest for future research with clinical approaches as antidiabetic agents.

In the short-term, ZnONPs supplementation could induce a hyperglycemic response by inhibition of insulin secretion. A report in β-cell islets showed that zinc inhibits insulin secretion concentration-dependent [39] and a recent study exhibited that zinc is a critical factor for synthesis and insulin secretion in β-cell [40]; ZnONPs could be dysregulating the insulin secretion pathway at any step in pancreatic β-cell. However, new studies are required to test the hypothesis here stated.

The differences observed in the temporal courses of hyperglycemic responses in healthy and diabetics can also be a result of a zinc homeostasis alterations, where capture and release are finely modulated to maintain a steady zinc concentration in the cells. The zinc transport is mainly performed by proteins controlling the influx (ZIP) and efflux (ZnT) [41]. There is evidence in the literature that shows differences in the gene expression of zinc transporters in healthy and diabetic rats. In diabetic rats a decrease in the expression of ZIP1 and ZIP4 is observed, which is associated with an over-expression of ZnT1, ZnT2, ZnT4, ZnT5, and ZnT7, which reduces the zinc bioavailability [42].

Pharmacokinetic data of the ZnONPs allows to better understand their effects and toxicity when are administered by different routes. In the experiments carried out in the present work, the hyperglycemic response induced by ZnONPs was lower when they were administered orally, compared to intraperitoneally route, the magnitude of this effect may be due to the fact that the absorption of ZnONPs is low through the gastrointestinal tract (6.5–32.5%), as has been demonstrated in previous studies [43]. On the other hand, it has been reported that clearance is higher when nanoparticles are orally administered in comparison with the intraperitoneal route. In fact, in oral administrations the maximum peak of zinc concentration in blood is reached six h after administration with a subsequent decrease to basal levels. Conversely, using the intraperitoneal route, the concentration is kept high 74 h after the administration, facilitating a greater biodistribution and accumulation of zinc in liver, spleen, lung, kidney and heart [44]. Furthermore, it has been shown that oral administration of ZnONPs generate a rapid clearance by defecation [43,44]. These differences in absorption and clearance could explain the reasons that ZnONPs intraperitoneally injected were more effective in our study. Although, further investigations are required.

Another interesting explanation for the hyperglycemic phenomenon observed in our experiments would be to study in mammals whether the ZnONPs induce post-administration an imbalance at the systemic level of the hormones that maintain glucose homeostasis, a decrease in insulin coupled with an increase in glucagon and cortisol could induce a hyperglycemic response, in this context there is scientific evidence in other species that shows that ZnONPs decrease the amount of insulin and increase glucagon and cortisol [45].

It is important to keep in mind in pharmacological research the dose ranges with no adverse or toxic effects, for this, the toxicity index called NOAEL (No-Observed-Adverse-Effect-Level) is used as an approximation. The NOAEL index for ZnONPs in rats is around 125 mg/kg [46]. In our study, we evaluated two doses below this range, 10 and 100 mg/kg because several reports argue that higher doses (250–2000 mg/kg) generate histological alterations in the liver [43,47]. Such alterations are directly related to the increase in ALT (Alanine aminotransferase) and AST (Aspartate aminotransferase) levels in studies carried out in mice [44,48]. Zinc is significantly absorbed in kidneys, which show damage in histopathological studies. However, creatinine and urea are not altered [48]. Is also known that body weight of rats administered with nanoparticles did not change with doses of 50 to 2000 mg/kg [43]. Bioaccumulation of zinc is also observed in pancreas, liver, and fatty tissue [49]. ZnONPs at doses of 10 mg/kg show no pathological or structural abnormalities in organs such as the liver, kidney, and pancreas [49]. Although, we tested with doses below the NOEL index, the results show that there are doses and routes of administration that can put the life of the diabetic at risk, the study also provides evidence that more detailed toxicology studies are required before thinking about its clinical use.

Finally, another important factor that may be responsible for the difference in the effects observed in this study between oral and intraperitoneal route of administration are the characteristics of the environment of ZnONPs. It has been reported that ZnONPs in gastric fluid increase their hydrodynamic size, and negative surface charge decreases, in contrast when they are placed in plasma ZnONPs decrease in size and increase negative surface charge [50].

In general, the results obtained in this research suggest that any treatment based on ZnONPs in diabetic patients should be taken with caution until an integral evaluation of the risk for adverse effects in future research is performed, including the risk of diabetic coma and compromise of life.

4. Conclusions

In the short-term, ZnONPs induce a hyperglycemic response in healthy and diabetic rats; the magnitude of the effect was dose and administration route-dependent. Besides, the hyperglycemic response was higher in diabetic animals. This study provides new information about of acute effects of ZnONPs on the circulating blood glucose levels that could limit its therapeutic application in diabetic patients. Nevertheless, future investigation is required to elucidate the mechanism of action of this compound.

To better understand the acute hyperglycemic effect induced by the ZnONPs, it would be essential to measure in a future study insulin, glucagon and cortisol levels in vivo after the administration of nanoparticles (0–6 h) and explore the effects on the liver metabolism. These experiments will allow us to know whether the ZnONPs act by an imbalance in hormones that regulate blood glucose or increasing hepatic glycogenolysis.

Author Contributions: Conceptualization, A.V.-O. and A.A.-I.; methodology, I.D.-R., H.P.-D. and S.L.-M.; formal analysis, I.D.-R., H.P.-D. and S.L.-M.; investigation, E.A.S.-P., S.J.C. and E.B.-A.; writing—original draft preparation, A.V.-O. and A.A.-I.; writing—review and editing, S.J.C., L.C.-S., A.R.-H. and A.D.-A.; supervision, A.V.-O. All authors have read and agreed to the published version of the manuscript.

Funding: This research was funded by Centro Universitario de Investigaciones Biomédicas, Universidad de Colima. The APC was funded by all authors.

Conflicts of Interest: The authors declare no conflict of interest.

References

1. Keller, A.A.; McFerran, S.; Lazareva, A.; Suh, S. Global life cycle releases of engineered nanomaterials. *J. Nanopart. Res.* **2013**, *15*, 1692. [CrossRef]
2. Wojnarowicz, J.; Chudoba, T.; Lojkowski, W. A Review of Microwave Synthesis of Zinc Oxide Nanomaterials: Reactants, Process Parameters and Morphoslogies. *Nanomaterials* **2020**, *10*, 1086. [CrossRef] [PubMed]
3. Napi, M.; Sultan, S.M.; Ismail, R.; How, K.W.; Ahmad, M.K. Electrochemical-Based Biosensors on Different Zinc Oxide Nanostructures: A Review. *Materials* **2019**, *12*, 2985. [CrossRef] [PubMed]
4. Zhang, Z.Y.; Xiong, H.M. Photoluminescent ZnO Nanoparticles and Their Biological Applications. *Materials* **2015**, *8*, 3101–3127. [CrossRef]
5. Patra, J.K.; Das, G.; Fraceto, L.F.; Campos, E.; Rodriguez-Torres, M.; Acosta-Torres, L.S.; Diaz-Torres, L.A.; Grillo, R.; Swamy, M.K.; Sharma, S.; et al. Nano based drug delivery systems: Recent developments and future prospects. *J. Nanobiotechnol.* **2018**, *16*, 71. [CrossRef]
6. Martínez-Carmona, M.; Gun'ko, Y.; Vallet-Regí, M. ZnO Nanostructures for Drug Delivery and Theranostic Applications. *Nanomaterials* **2018**, *8*, 268. [CrossRef]
7. Bisht, G.; Rayamajhi, S. ZnO Nanoparticles: A Promising Anticancer Agent. *Nanobiomedicine* **2016**, *3*, 9. [CrossRef]
8. Alkaladi, A.; Abdelazim, A.M.; Afifi, M. Antidiabetic activity of zinc oxide and silver nanoparticles on streptozotocin-induced diabetic rats. *Int. J. Mol. Sci.* **2014**, *15*, 2015–2023. [CrossRef]
9. Lallo da Silva, B.; Abuçafy, M.P.; Berbel Manaia, E.; Oshiro Junior, J.A.; Chiari-Andréo, B.G.; Pietro, R.; Chiavacci, L.A. Relationship Between Structure And Antimicrobial Activity Of Zinc Oxide Nanoparticles: An Overview. *Int. J. Nanomed.* **2019**, *14*, 9395–9410. [CrossRef]
10. Liang, H.; Chen, A.; Lai, X.; Liu, J.; Wu, J.; Kang, Y.; Wang, X.; Shao, L. Neuroinflammation is induced by tongue-instilled ZnO nanoparticles via the Ca^{2+}-dependent NF-κB and MAPK pathways. *Part. Fibre Toxicol.* **2018**, *15*, 39. [CrossRef]

11. Virgen-Ortiz, A.; Limón-Miranda, S.; Soto-Covarrubias, M.A.; Apolinar-Iribe, A.; Rodríguez-León, E.; Iñiguez-Palomares, R. Biocompatible silver nanoparticles synthesized using rumex hymenosepalus extract decreases fasting glucose levels in diabetic rats. *Dig. J. Nanomater. Biostruct.* **2015**, *10*, 927–933.
12. Shaheen, T.I.; El-Naggar, M.E.; Hussein, J.S.; El-Bana, M.; Emara, E.; El-Khayat, Z.; Fouda, M.M.G.; Ebaid, H.; Hebeish, A. Antidiabetic assessment; in vivo study of gold and core-shell silver-gold nanoparticles on streptozotocin-induced diabetic rats. *Biomed. Pharmacother.* **2016**, *83*, 865–875. [CrossRef] [PubMed]
13. Opris, R.; Tatomir, C.; Olteanu, D.; Moldovan, R.; Moldovan, B.; David, L.; Nagy, A.; Decea, N.; Kiss, M.L.; Filip, G.A. The effect of Sambucus nigra L. extract and phytosinthesized gold nanoparticles on diabetic rats. *Colloids Surf. B.* **2017**, *150*, 192–200. [CrossRef] [PubMed]
14. Sharma, A.K.; Kumar, A.; Taneja, G.; Nagaich, U.; Deep, A.; Rajput, S.K. Synthesis and preliminary therapeutic evaluation of cooper nanoparticles against diabetes mellitus and induced micro-(renal) and macro-(vascular (vascular endothelial and cardiovascular) abnormalities in rats. *RSC Adv.* **2016**, *6*, 36870–36880. [CrossRef]
15. Ahmed, H.H.; Abd El-Maksoud, M.D.; Abdel Moneim, A.E.; Aglan, H.A. Pre-clinical study for the antidiabetic potencial of selenium nanoparticles. *Biol. Trace Elem. Res.* **2017**, *177*, 267–280. [CrossRef]
16. Naghsh, N.; Kazemi, S. Effect of nano-magnesium oxide on glucose concentration and lipid profile in diabetic laboratory mice. *Iran J. Pharm. Sci.* **2014**, *10*, 63–68.
17. Zhai, J.H.; Wu, Y.; Wang, X.Y.; Cao, Y.; Xu, K.; Xu, L.; Guo, Y. Antioxidation of Cerium Oxide Nanoparticles to Several Series of Oxidative Damage Related to Type II Diabetes Mellitus In Vitro. *Med. Sci. Monit.* **2016**, *22*, 3792–3797. [CrossRef]
18. Ziental, D.; Czarczynska-Goslinska, B.; Mlynarczyk, D.T.; Glowacka-Sobotta, A.; Stanisz, B.; Goslinski, T.; Sobotta, L. Titanium Dioxide Nanoparticles: Prospects and Applications in Medicine. *Nanomaterials* **2020**, *10*, 387. [CrossRef]
19. Ahmed, F.; Husain, Q.; Ansari, M.O.; Shadab, G.G.H.A. Antidiabetic and oxidative stress assessment of bio-enzymatically synthesized zinc oxide nanoformulation on streptozotocin-induced hyperglycemic mic. *Appl. Nanosci.* **2020**, *10*, 879–893. [CrossRef]
20. Prabhu, S.; Vinodhini, S.; Elanchezhiyan, C.; Rajeswari, D. Evaluation of antidiabetic activity of biologically synthesized silver nanoparticles using Pouteria sapota in streptozotocin-induced diabetic rats. *J. Diabetes* **2018**, *10*, 28–42. [CrossRef]
21. Bayrami, A.; Parvinroo, S.; Habibi-Yangjeh, A.; Rahim-Pouran, S. Bio-extract-mediated ZnO nanoparticles: Microwave-assisted synthesis, characterization and antidiabetic activity evaluation. *Artif. Cells Nanomed. Biotechnol.* **2018**, *46*, 730–739. [CrossRef] [PubMed]
22. El-Gharbawy, R.M.; Emara, A.M.; Abu-Risha, S.E. Zinc oxide nanoparticles and a standard antidiabetic drug restore the function and structure of beta cells in Type-2 diabetes. *Biomed. Pharmacother.* **2016**, *84*, 810–820. [CrossRef] [PubMed]
23. Nazarizadeh, A.; Asri-Rezaie, S. Comparative Study of Antidiabetic Activity and Oxidative Stress Induced by Zinc Oxide Nanoparticles and Zinc Sulfate in Diabetic Rats. *AAPS Pharm. Sci. Tech.* **2016**, *17*, 834–843. [CrossRef]
24. Desouza, C.; Salazar, H.; Cheong, B.; Murgo, J.; Fonseca, V. Association of hypoglycemia and cardiac ischemia: A study based on continuous monitoring. *Diabetes Care.* **2003**, *26*, 1485–1489. [CrossRef] [PubMed]
25. Beckman, J.A.; Creager, M.A.; Libby, P. Diabetes and atherosclerosis: Epidemiology, pathophysiology, and management. *JAMA* **2002**, *287*, 2570–2581. [CrossRef]
26. Gogitidze-Joy, N.; Hedrington, M.S.; Briscoe, V.J.; Tate, D.B.; Ertl, A.C.; Davis, S.N. Effects of acute hypoglycemia on inflammatory and pro-atherothrombotic biomarkers in individuals with type 1 diabetes and healthy individuals. *Diabetes Care* **2010**, *33*, 2129. [CrossRef]
27. Kawahito, S.; Kitahata, H.; Oshita, S. Problems associated with glucose toxicity: Role of hyperglycemia-induced oxidative stress. *World J. Gastroenterol.* **2009**, *15*, 4137–4142. [CrossRef]
28. Maiga, D.; Nyoni, H.; Nkambule, T.; Mamba, B.; Msagati, T. Impact of zinc oxide nanoparticles in aqueous environments: Influence of concentrations, natural organic matter and ionic strength. *Inorg. Nano-Met. Chem.* **2020**, *50*, 680–692. [CrossRef]
29. Wu, W.; Ichihara, G.; Suzuki, Y.; Izuoka, K.; Oikawa-Tada, S.; Chang, J.; Sakai, K.; Miyazawa, K.; Porter, D.; Castranova, V.; et al. Dispersion method for safety research on manufactured nanomaterials. *Ind. Health* **2014**, *52*, 54–65. [CrossRef]

30. Masiello, P.; De Paoli, A.A.; Bergamini, E. Influence of age on the sensitivity of the rat to streptozotocin. *Horm Res.* **1979**, *11*, 262–274. [CrossRef]
31. Masiello, P.; Broca, C.; Gross, R.; Roye, M.; Manteghetti, M.; Hillaire-Buys, D.; Novelli, M.; Ribes, G. Experimental NIDDM: Development of a new model in adult rats administered streptozotocin and nicotinamide. *Diabetes* **1998**, *47*, 224–229. [CrossRef] [PubMed]
32. Wahba, N.S.; Shaban, S.F.; Kattaia, A.A.A.; Kandeel, S.A. Efficacy of zinc oxide nanoparticles in attenuating pancreatic damage in a rat model of streptozotocin-induced diabetes. *Ultrastruct. Pathol.* **2016**, *40*, 358–373. [CrossRef] [PubMed]
33. Amiri, A.; Dehkordi, R.A.F.; Heidarnejad, M.S.; Dehkordi, M.J. Effect of the Zinc Oxide Nanoparticles and Thiamine for the Management of Diabetes in Alloxan-Induced Mice: A Stereological and Biochemical Study. *Biol. Trace Elem. Res.* **2018**, *181*, 258–264. [CrossRef] [PubMed]
34. Rehana, D.; Mahendiran, D.; Kumar, R.S.; Rahiman, A.K. In vitro antioxidant and antidiabetic activities of zinc oxide nanoparticles synthesized using different plant extracts. *Bioprocess Biosyst. Eng.* **2017**, *40*, 943–957. [CrossRef] [PubMed]
35. Asani, S.C.; Umrani, R.D.; Paknikar, K.M. Differential dose-dependent effects of zinc oxide nanoparticles on oxidative stress-mediated pancreatic β-cell death. *Nanomedicine* **2017**, *12*, 745–759. [CrossRef]
36. Nathan, D.M.; Buse, J.B.; Davidson, M.B.; Ferrannini, E.; Holman, R.R.; Sherwin, R.; Zinman, B. Medical management of hyperglycaemia in type 2 diabetes mellitus: A consensus algorithm for the initiation and adjustment of therapy: A consensus statement from the American Diabetes Association and the European Association for the Study of Diabetes. *Diabetologia* **2009**, *52*, 17–30. [CrossRef]
37. Brand, I.A.; Kleineke, J. Intracellular zinc movement and its effect on the carbohydrate metabolism of isolated rat hepatocytes. *J. Biol. Chem.* **1996**, *271*, 1941–1949. [CrossRef]
38. Kenneth, R.E.; Robert, J.C. Hyperglycemic action of zinc in rats. *J. Nutr.* **1983**, *113*, 1657–1663.
39. Ghafghazi, T.; McDaniel, M.L.; Lacy, P.E. Zinc-induced inhibition of insulin secretion from isolated rat islets of Langerhans. *Diabetes* **1981**, *30*, 341–345. [CrossRef]
40. Fukunaka, A.; Fujitani, Y. Role of Zinc Homeostasis in the Pathogenesis of Diabetes and Obesity. *Int. J. Mol. Sci.* **2018**, *19*, 476. [CrossRef]
41. Hara, T.; Takeda, T.; Takagishi, T.; Fukue, K.; Kambe, T.; Fukada, T. Physiological roles of zinc transporters: Molecular and genetic importance in zinc homeostasis. *J. Physiol. Sci.* **2017**, *67*, 283–301. [CrossRef] [PubMed]
42. Barman, S.; Pradeep, S.R.; Srinivasan, K. Zinc supplementation mitigates its dyshomeostasis in experimental diabetic rats by regulating the expression of zinc transporters and metallothionein. *Metallomics* **2017**, *9*, 1765–1777. [CrossRef] [PubMed]
43. Baek, M.; Chung, H.E.; Yu, J.; Lee, J.A.; Kim, T.H.; Oh, J.M.; Lee, W.J.; Paek, S.M.; Lee, J.K.; Jeong, J. Pharmacokinetics, tissue distribution, and excretion of zinc oxide nanoparticles. *Int. J. Nanomed.* **2012**, *7*, 3081–3097.
44. Li, C.-H.; Shen, C.-C.; Cheng, Y.-W.; Huang, S.-H.; Wu, C.-C.; Kao, C.-C.; Liao, J.-W.; Kang, J.-J. Organ biodistribution, clearance, and genotoxicity of orally administered zinc oxide nanoparticles in mice. *Nanotoxicology* **2012**, *6*, 746–756. [CrossRef] [PubMed]
45. Alkaladi, A.; Afifi, M.; Ali, H.; Saddick, S. Hormonal and molecular alterations induced by sub-lethal toxicity of zinc oxide nanoparticles on Oreochromis niloticus. *Saudi J. Biol. Sci.* **2020**, *27*, 1296–1301. [CrossRef] [PubMed]
46. Chung, H.E.; Yu, J.; Baek, M.; Lee, J.A.; Kim, M.S.; Kim, S.H.; Maeng, S.; Lee, J.K.; Jeong, J.; Choi, S.J. Toxicokinetics of zinc oxide nanoparticles in rats. *J. Phys. Conf. Ser.* **2013**, *429*, 012037. [CrossRef]
47. Ansar, S.; Abudawood, M.; Alaraj, A.S.A.; Hamed, S.S. Hesperidin alleviates zinc oxide nanoparticle induced hepatotoxicity and oxidative stress. *BMC Pharmacol. Toxicol.* **2018**, *19*, 65. [CrossRef]
48. Esmaeillou, M.; Moharamnejad, M.; Hsankhani, R.; Tehrani, A.A.; Maadi, H. Toxicity of nanoparticles in healthy adult mice. *Environ. Toxicol. Pharmacol.* **2013**, *35*, 67–71. [CrossRef]

49. Umrani, R.D.; Paknikar, K.M. Zinc oxide nanoparticles show antidiabetic activity in streptozotocin-induced Type 1 and 2 diabetic rats. *Nanomedicine* **2014**, *9*, 89–104. [CrossRef]
50. Yu, J.; Kim, H.J.; Go, M.R.; Bae, S.H.; Choi, S.J. ZnO Interactions with Biomatrices: Effect of Particle Size on ZnO-Protein Corona. *Nanomaterials* **2017**, *7*, 377. [CrossRef]

Article

Evaluation of the NLRP3 Inflammasome Activating Effects of a Large Panel of TiO_2 Nanomaterials in Macrophages

Julia Kolling [1], Jonas Tigges [1,2], Bryan Hellack [3,4], Catrin Albrecht [1,5] and Roel P. F. Schins [1,*]

1 IUF—Leibniz Research Institute for Environmental Medicine, Auf'm Hennekamp 50, 40225 DE Düsseldorf, Germany; julia.kolling@gmx.de (J.K.); jonas.tigges@gmx.de (J.T.); catrin.albrecht@sachsen-anhalt.de (C.A.)
2 Bundeswehr Institute of Pharmacology and Toxicology, Neuherbergstraße 11, 80937 DE Munich, Germany
3 Institute of Energy and Environmental Technology e.V. (IUTA), Bliersheimer Str. 60, 47229 Duisburg, Germany; bryan.hellack@uba.de
4 UBA—German Environment Agency, Paul-Ehrlich-Str. 29, 63225 Langen, Germany
5 State Office for Consumer Protection Saxony-Anhalt, 39576 Stendal, Germany
* Correspondence: roel.schins@uni-duesseldorf.de; Tel.: +49-211-3389-269

Received: 16 July 2020; Accepted: 15 September 2020; Published: 19 September 2020

Abstract: TiO_2 nanomaterials are among the most commonly produced and used engineered nanomaterials (NMs) in the world. There is controversy regarding their ability to induce inflammation-mediated lung injuries following inhalation exposure. Activation of the NACHT, LRR and PYD domains-containing protein 3 (NALP3) inflammasome and subsequent release of the cytokine interleukin (IL)-1β in pulmonary macrophages has been postulated as an essential pathway for the inflammatory and associated tissue-remodeling effects of toxic particles. Our study aim was to determine and rank the IL-1β activating properties of TiO_2 NMs by comparing a large panel of different samples against each other as well as against fine TiO_2, synthetic amorphous silica and crystalline silica (DQ12 quartz). Effects were evaluated in primary bone marrow derived macrophages (BMDMs) from NALP3-deficient and proficient mice as well as in the rat alveolar macrophage cell line NR8383. Our results show that specific TiO_2 NMs can activate the inflammasome in macrophages albeit with a markedly lower potency than amorphous SiO_2 and quartz. The heterogeneity in IL-1β release observed in our study among 19 different TiO_2 NMs underscores the relevance of case-by-case evaluation of nanomaterials of similar chemical composition. Our findings also further promote the NR8383 cell line as a promising in vitro tool for the assessment of the inflammatory and inflammasome activating properties of NMs.

Keywords: nanomaterials; titanium dioxide; NALP3; interleukin-1beta; NR8383; bone marrow derived macrophages

---

1. Introduction

Due to their unique properties, engineered nanomaterials (NMs) have been used since many decades in several different applications. With their increasing production and potential exposure, there is also rising concern about possible harmful properties of these compounds regarding human health. This is also the case for titanium dioxide (TiO_2) NMs; they are among the most produced nanomaterials worldwide and are applied in a large variety of sectors including agriculture, energy, the food and cosmetic industries as well as in chemical and biomedical research [1,2].

Inhalation of crystalline silica is well known to trigger lung inflammation. And with high and persistent exposures to these mineral dust particles, the resulting sustained inflammation is implicated in the development of deliberating lung diseases including silicosis and cancer [3]. The inflammatory effects of crystalline silica particles are driven by their surface chemistry and ability to generate reactive oxygen species (ROS) and associated oxidative stress [4,5]. Compared to crystalline silica, TiO_2 particles are traditionally thought to have a relatively low toxicity. Yet, there is concern for possible adverse health effects of TiO_2 NMs, which has been linked to ROS generation and inflammatory processes upon inhalation as well [6–8]. It has been shown that particle size and surface are important characteristics of TiO_2 with particles in a nano-scale (20 nm) being much more cytotoxic than fine TiO_2 (250 nm), driven by their increased specific surface area dose [9,10]. Other acute inhalation studies detected microvascular dysfunctions and peripheral vascular effects with nano TiO_2 being six to seven times more reactive than fine TiO_2 [11]. Normalization of the dose on particle surface area basis resulted in an equal potency for fine- and nano-TiO_2.

With the identification of the "NACHT, LRR and PYD domains-containing protein 3" (NALP3) inflammasome as an essential component in crystalline silica induced lung inflammation and fibrosis [12–14] it has also been proposed that this pathway dominates the inflammatory properties of inhaled nanomaterials. The NALP3 inflammasome is a central activator of the innate immune defense in response to cellular infections and is capable of cleaving pro caspase-1 obtaining the biological active caspase-1. Caspase-1 in its active form cleaves the inactive pro Interleukin-1β (IL-1β), generating mature IL-1β [15]. It has been suggested, that NMs cause lysosomal rupture upon particle phagocytosis leading to ROS release into the cytoplasm and subsequent NALP3 inflammasome activation [16–18]. Therefore, IL-1β, as product of NALP3 inflammasome activation, can be seen as marker for particle induced inflammation.

In crystalline silica-exposed lungs, macrophages have emerged as key players in NALP3-mediated inflammation and tissue remodeling, although a contribution of structural epithelial cells cannot be excluded [19]. In the case of TiO_2 NMs, the literature provides contrasting data regarding inflammasome activation in professional phagocytes of the innate immune system. Whereas some research groups detected an increased secretion of IL-1β from TiO_2 treated bone marrow derived macrophages (BMDMs) [20,21], human THP-1 macrophage-like cells [8,22] or bone marrow derived dendritic cells (BMDCs) [23], other studies did not indicate an up-regulated release following TiO_2 treatment in mouse RAW 264.7 cells [24], NR8383 rat alveolar macrophages [25] or BMDMs [26]. At least to some extent, these differences may be explained by differences in the cell-type used, but they could also relate to the selected type and even batch of the TiO_2 nanomaterial as well as the method of their application to the cell system.

The aim of our study was, therefore, to determine to what extent TiO_2 NMs are capable of activating the NALP3 inflammasome in macrophages. We selected a large panel of different TiO_2 NMs and compared their effects against each other as well as against a sample of fine TiO_2, a synthetic amorphous SiO_2 and the well-investigated crystalline silica sample DQ12. The investigations were performed with BMDMs obtained from NALP3-deficient and NALP3 proficient mice as well in the well-established rat alveolar macrophages cell line NR8383. Our study findings are discussed in relation to current debate on the toxicity of NMs, specifically, regarding the inflammatory and inflammasome activating properties of TiO_2 NMs in the lung.

2. Materials and Methods

2.1. Particles

For this study we selected a total of 19 different TiO_2 NMs (abbreviated in this study as NT1 to NT19). The selected NMs have been the subject of investigation in various nanosafety and metrology projects. Our main experiments were performed with four TiO_2 NMs, referred to as NT1 to NT4. Their origin and pristine characteristics are listed in Table 1. The samples NT1 to NT3 are three commercial samples, characterized by nearly spherical primary nanoparticles. Sample NT4 represents a truncated bipyramid shape TiO_2 NM with an aspect ratio of 3:2, which has been synthesized by the University of Turin (Turin, Italy). Together with seven further test samples (i.e., NT5 to NT11) these NMs were all purchased or synthesized, and subsequently characterized, within the framework of the EU FP7 metrology project SetNanoMetro [27–29]. The other TiO_2 NMs used in our study included the JRC repository samples NM100 (=NT12), NM101 (=NT13), NM103 (=NT14) and NM104 (=NT15) characterized and previously studied within the SIINN ERANET project NanOxiMet (for characteristics, see: https://www.nanopartikel.info/projekte/era-net-siinn/nanoximet/) and a set of TiO_2 NMs that was characterized and investigated within the FP7 project ENPRA, i.e., NM101 (=N16), NRCWE001 (=NT17), NRCWE002 (=NT18) and NRCWE 003 (=NT19) (for characteristics, see: [30,31]). As negative control, a well-investigated fine TiO_2 (=FT) was included [23,25]. Finally, amorphous silica (=AS) and fine crystalline silica (CS) were used as well-established positive controls [13,17,32]. The FT was obtained from Sigma Aldrich, Taufkirchen, Germany and represents a pure anatase sample with a BET surface area of 10 m^2/g and a reported mean diameter of about 250 nm. The AS was purchased from Sigma Aldrich (#S5130), Taufkirchen, Germany, a fumed silica with a declared primary particle size of 7 nm and mean surface area of 395 ± 25 m^2/g. The CS sample (DQ12 quartz, batch 6, IUF) has a mean particle size of 960 ± 620 nm, a surface area of 9.6 m^2/g and a quartz content of 87% [33].

Table 1. Characteristics of the pristine TiO_2 nanomaterials.

Sample	ID	Supplier	Diameter [1] (nm)	BET [2] (m^2/g)	Crystal Phase [3]	Method of Synthesis
NT1	P25	Evonik	12–18	55	A/R	flame pyrolysis of $TiCl_4$
NT2	PC105	Cristal	10	86	A	hydrolysis of titanyl sulfate and unspecified thermal treatment
NT3	SX001	Solaronix	12–15	93	A	hydrothermal process
NT4	UT001	UNITO [4]	16–17	47	A	hydrolysis of aqueous solution of Ti^{IV}(triethanolamine)$_2$titanatrane

[1] Primary particle size; [2] Specific surface area according to Brunauer Emmett and Teller; [3] A = Anatase/R = Rutile; [4] University of Turin.

2.2. *Isolation and Differentiation of Bone Marrow Derived Macrophages*

The NALP3 inflammasome activating properties of the TiO_2 NMs in macrophages were investigated using BMDM from mice. Therefore, C57BL/6J mice, originally purchased from Jackson Laboratories, Bar Harbor, ME, USA, were obtained from in-house breeding. In addition, B6.129S6-Nlrp3tm1Bhk/J mice were purchased from Jackson Laboratories [34]. For the preparation of BMDMs, mice with an age of 4 to 10 months were used. The mice were maintained according to the guidelines of the Society for Laboratory Animals Science (GV-SOLAS). The experiments (i.e., organ removal) were approved by the State Office for Nature, Environment and Consumer Protection of North Rhine-Westphalia, Germany (Landesamt für Natur, Umwelt und Verbraucherschutz, LANUV reference 84-02.05.40.14.138). The mice were sacrificed either by cervical dislocation or via i.p. injection of Phenobarbital (Narcoren®, Merial GmbH, Hallbergmoos, Germany, 800 mg/kg b.w.). Both hind legs were amputated, muscles as well as connective tissues were removed and the femur was cut and flushed with cold PBS from distal to proximal. Bone marrow of both femurs was combined and dispersed until cells were sufficiently separated. Afterwards, cells were passed through a 100 μm cell strainer to separate debris. The strainer was washed with PBS and cells were centrifuged at 800× *g* for 5 min. The cell pellet was re-suspended in 10 mL complete differentiation medium, consisting of RPMI 1640 Medium, containing 10% FCS, 10% L929 supernatant, 2% glutamine, 1% Penicillin/Streptomycin and 0.1% β-mercaptoethanol. Next, 5 × 10^6 cells in 10 mL culture medium were seeded into a 100 mm bacteria culture dish and incubated at 37 °C and 5% CO_2. After three days, 10 mL complete differentiation medium were added to each culture dish. At day six, adherent fraction represented the differentiated macrophages. To determine the success of the differentiation process, cells were stained with antibodies against F4/80 and CD11b and were analyzed by fluorescence activated cell sorting (FACS). Viability of all analyzed cells was between 77.2% and 95.8% and the percentage of differentiated macrophages reached from 77.5% to 97%. One day prior to each experiment, cells were seeded in a concentration of 5 × 10^4 cells/cm^2 in 96-well plates and incubated at 37 °C and 5% CO_2.

2.3. NR8383 Cells

The NR8383 rat alveolar macrophage cell line, obtained from ATCC ((CRL-2192), Manassas, VA, USA) was cultured in DMEM/F-12 medium (Thermo Fisher Scientific, Waltham, MA, USA) containing 15% FCS, 1% penicillin/streptomycin and 1% glutamine (all purchased from Sigma-Aldrich, Taufkirchen, Germany) and incubated at 37 °C and 5% CO_2. Two days prior to each experiment, cells were seeded at a density of 4×10^4 cells/cm^2 in 96-well plates and incubated at 37 °C and 5% CO_2.

2.4. Treatment of Cells

The particle suspensions were prepared by dispersion in HLPC grade water at a concentration of 2 mg/mL and then sonicated with a Cuphorn (Branson Sonifier 450, Brookfield, CT, USA) for 10 min (Duty cycle 20%, power 5.71 (200 W)). Final particle concentrations were achieved by dilution of the particle suspensions with cell culture medium. The NR8383 cells were treated 48 h after seeding with particles at the indicated concentrations in medium without serum and phenol red whereas the BMDMs were treated in complete (i.e., FCS containing) medium 24 h after seeding. The experiments in the NR8383 cells were performed in the absence of FCS to abrogate proliferation of this immortalized cell line. As such, this in vitro model better reflects the typical non-proliferative phenotype of the resident macrophages of the lung alveoli. Moreover, it avoids a dose dilution over treatment time as would occur with proliferating cells in terms particle mass (or number) per unit cell number.

2.5. Dynamic Light Scattering

The dispersion states of the NMs in cell culture media were evaluated by dynamic light scattering (DLS) using a Delsa-Nano C (Beckman Coulter Inc., Krefeld, Germany). The measurements were performed for three types of suspensions relating to the specific dispersion protocol (including sonication and the respective culture media used for the NR8383 cells and BMDMs). Cumulative diameter and polydispersity index were determined for the four used TiO_2 NMs (NT1-NT4) as well as the synthetic amorphous silica (AS), after suspension in dH_2O or the DMEM and RPMI medium used for experiments. Results of these measurements are shown in Table 2, and represent mean values of three independent experiments.

Table 2. Nanoparticle characteristics by Dynamic Light Scattering.

	Cumulant Diameter [1]			Polydispersity Index [1]		
Sample	**dH_2O**	**DMEM**	**RPMI**	**dH_2O**	**DMEM**	**RPMI**
NT1	193 ± 11	2338 ± 55	250 ± 5	0.208 ± 0.03	0.467 ± 0.02	0.191 ± 0.01
NT2	650 ± 19	1188 ± 28	775 ± 17	0.222 ± 0.07	0.380 ± 0.004	0.275 ± 0.01
NT3	516 ± 118	1684 ± 73	476 ± 118	0.235 ± 0.05	0.530 ± 0.03	0.215 ± 0.05
NT4	207 ± 18	1987 ± 50	335 ± 44	0.135 ± 0.03	0.456 ± 0.02	0.154 ± 0.02
AS	181 ± 1.1	192 ± 2.7	338 ± 4.3	0.144 ± 0.02	0.174 ± 0.01	0.257 ± 0.02

[1] Data represent mean ± SD (n = 3).

2.6. WST-1 Assay

Cell viability was assessed using the WST-1 assay (Roche Diagnostics GmbH, Mannheim, Germany). Following 4 h or 24 h of particle treatment, two out of six replicates of each test condition were additionally treated with 1% Triton-X for 5 min, which served as positive control for maximal cell death and particle absorption. After addition of 10 µL WST-1 solution per well, cells were incubated for 2 h at 37 °C and 5% CO_2. Afterwards, optical density was detected at 450 nm and 630 nm using a Thermo Multiskan GO Microplate Spectrophotometer (Thermo Fisher Scientific, Waltham, MA, USA) and percentage of mitochondrial activity related to the control was calculated. Particle absorption in cell-free samples was detected and subtracted from the calculated values of particle treated cells to exclude particle related effects.

2.7. IL-1β ELISA

For the assessment of inflammasome activation, 5 × 10^5 cells per well of a 96 well plate were pre-treated with 10 (BMDMs) or 100 (NR8383) ng/mL lipopolysaccharide (LPS) for 4 h, to induce pro-IL1 β transcription. The respective priming concentrations of the LPS were selected on the basis of pilot experiments using CS (data not shown). Following LPS-priming, the cells were exposed to the different particles for 4 h or 24 h at the indicated concentrations. Cell free cell culture supernatants were collected and the amount of secreted IL-1β was detected by ELISA on a Thermo Multiskan GO Microplate Spectrophotometer (Thermo Fisher Scientific, Waltham, MA, USA), using commercial detection kits for mouse (i.e., BMDMs) (Bio-Techne Corporation, Minneapolis, MN, USA) or rat (i.e., NR8383 cells) (R&D Systems #RLB00).

2.8. mRNA Expression Analyses by qRT-PCR

NR8383 cells were seeded in 6-well plates, treated with particles for 4 h, scraped and centrifuged (200 g, 5 min, 4 °C). The pellet was resuspended in 0.5 mL Trizol® Reagent (Invitrogen GmbH, Karlsruhe, Germany) and stored at −80 °C until further use. For RNA extraction, 200 μL chloroform were added to each sample and incubated for 3 min at RT followed by centrifugation at 12,000 rcf at 4 °C for 15 min to separate the phases. The aqueous phase was transferred to a new tube and 400 μL Isopropanol were added. Samples were incubated for 10 min at room temperature followed by centrifugation at 12,000 rcf and 4 °C for 15 min. The RNA pellet was further washed using 75% ethanol and centrifugation for 5 min at 7500 rcf. RNA pellet was air dried and re-suspended in RNase-free water. Finally, samples were incubated for 10 min at 60 °C and purity of RNA was evaluated using spectrophotometry at 260 and 280 nm. cDNA was synthesized using the iScript™ cDNA Synthesis kit (BioRad, Hercules, CA, USA. cDNA was diluted 15× in RNAse-free water before use. Primer sequences for Heme oxygenase-1 (HO-1) were 5′-GGG AAG GCC TGG CTT TTTT -3′ (forward) and 5′-CAC GAT AGA GCT GTT TGA ACT TGGT -3′ (reverse), for inducible Nitric Oxide Synthase (iNOS) 5′- AGG AGA GAG ATC CGG TTC ACA GT -3′ (forward) and 5′- ACC TTC CGC ATT AGC ACA GAA -3′ (reverse), for IL-1β 5′-CAG GAA GGC AGT GTC ACT CA-3′ (forward) and 5′-AAA GAA GGT GCT TGG GTC CT -3′ (reverse), for IL-6 5′-GCC CTT CAG GAA CAG CTA TGA-3′ (forward) and 5′-TGT CAA CAA CAT CAG TCC CAA GA-3′ (reverse) and for β-actin 5′-CCC TGG CTC CTA GCA CCA T-3′ (forward) and 5′-ATA GAG CCA CCA ATC CAC ACA GA-3′ (reverse). qRT-PCR was performed with a MyiQ Single Color real time PCR detection system (BioRad, Hercules, CA, USA) using iQ™ SYBR® Green Supermix (Biorad), 5 μL diluted cDNA, and 2.5 μL of 0.3 μM forward and reverse primer in a total volume of 20 μL. PCR was conducted as follows: a denaturation step at 95 °C for 3 min was followed by 40 cycles at 95 °C (15 s) and 60 °C (45 s). After PCR, a melt curve (55–95 °C) was generated for product identification and purity. Data were analyzed using the MyiQ Software system (BioRad) and were expressed as relative gene expression (fold increase) using the $2^{-\Delta\Delta Ct}$ method of [35] with β-actin as house-keeping gene.

2.9. ROS Measurement by Electron Paramagnetic Resonance (EPR) Spectroscopy

The amount of ROS formed in the treated BMDMs was evaluated using EPR spectroscopy with the use of the spin-trapping compound 5,5-dimethylpyrroline N-oxide (DMPO). Therefore, 31.25 × 10^5 BMDMs per well of a 24 well plate were either primed for 4 h with LPS (10 ng/mL) or left un-primed. After 4 h, treatment medium was replaced by 100 μL Hank's Balanced Salt Solution (HBSS) $^{(+/+)}$. Thereafter, 40 μg/cm^2 of particles or 6 mM H_2O_2 (as positive control) were added, followed by the addition of 0.1 M DMPO (Sigma-Aldrich, Taufkirchen, Germany). After 1 h or 3 h incubation at 37 °C and 5% CO_2, cell-free supernatants were harvested and immediately measured for radical formation using a MiniScope MS200 Spectrometer (Magnettech, Berlin, Germany) as described in [25]. Quantification was carried out on first derivation of EPR signal of the characteristic DMPO-OH quartet, as the mean of amplitudes, and outcomes are expressed in arbitrary units (a.u.).

2.10. Statistical Analyses

Statistical significances of experimental results were calculated by one-way analysis of variance (ANOVA) followed by Dunnett's multiple comparison test for comparison of multiple treatments to the control. Student's t-test was used for detection of significant differences between knock out and wild type macrophages. Significance was ascribed at $p < 0.05$. Analyses were conducted using SPSS statistics, Version 22 (IBM Corporation, Armonk, NY, USA).

3. Results

3.1. Effects of TiO_2 NMs in BMDMs

The ability of NMs to induce maturation and subsequent release of IL-1β, as detected by ELISA, is in support of their ability to activate the inflammasome pathway. Therefore, the BMDMs were pre-stimulated with LPS to induce transcription of the pro-form of IL-1β and subsequently treated with the various particles [23]. First, the effects of these treatment conditions on cellular viability were evaluated. BMDMs obtained from C57BL/6J mice were treated for 4 h with 10 ng/mL LPS and afterwards treated for 4 h and 24 h with 5 to 40 μg/cm^2 of the particles (see Figure 1A,D). After 4 h treatment, only the viability of cells treated with the amorphous SiO_2 (AS) or crystalline silica (CS) was significantly decreased. None of the four TiO_2 NMs caused a decrease in cellular viability for this treatment time interval. After 24 h, however, viability was also significantly decreased by both NT1 and NT4 at the highest treatment concentrations (40 μg/cm^2). The viability of the BMBMs was further decreased by AS and CS at 24 h compared to 4 h of treatment.

The release of IL-1β from the LPS pre-activated BMDMs by the different TiO_2, AS and CS particles was then evaluated. Results are shown in Figure S1 (Supplementary Materials). Clear dose-dependent increases in IL-1β release were found for all investigated particles after 24 h treatment, suggestive of their ability to activate the NALP3 inflammasome in macrophages. However, large differences in IL-1β release were observed for the different particle types. The strongest responses were observed with AS and CS. The TiO_2 particles showed much lower responses. Among them, NT2 revealed a markedly stronger effect than the other three TiO_2 NMs. NT3 appeared the least active nanomaterial.

Importantly, even after adjustment of the differences in surface area, the differences in potency of the TiO_2 NMs versus the amorphous SiO_2 remained obvious. With a BET of 86 m^2/g, the specific surface area of the NT2 is 9-fold higher than that of the CS (9.6 m^2/g) and 5-fold lower than that of the AS (395 m^2/g). Thus, when expressing the dose as BET surface area per cell culture dish area, the AS still caused an at least 2.5-fold higher IL-1β release than NT2. Similarly, this also demonstrated a much stronger IL-1β releasing response for CS compared to the amorphous AS.

To further explore if TiO_2 NMs activate the inflammasome and to verify if they act differently in this activation, the release of IL-1β was then compared using BMBMs from NALP3 deficient and NALP3 proficient mice. Supernatants from BMDMs from both backgrounds were therefore analyzed in parallel after 4 h and 24 h treatment with all particles at equal mass dose (i.e., 40 μg/cm^2). Results are shown in Figure 1. In the absence of LPS priming, no IL-1β release was detectable, neither in controls, nor in the particles treated BMDMs (data not shown). Using the LPS-priming protocol, IL-1β concentrations were increased for all four TiO_2 NMs after 4 h, although this was significant only for NT2 (Figure 1B). The increase after treatment with NT3 was much lower than after treatment with NT2, confirming strong differences between the TiO_2 NMs. Furthermore, 4 h treatment with AS and CS led to a much stronger increase of IL-1β secretion, which was about 5 times higher than for NT2 (Figure 1C). After 24 h, concentrations of IL-1β further increased and were significant for NT1, NT2 and NT4 (Figure 1E). Also, at this time, differences in IL-1β stimulating properties remained obvious among the four TiO_2 samples, as well as in comparison to the AS and CS (Figure 1F).

Using NALP3 deficient cells IL-1β release after 4 h was only detectable for NT2, and at a higher level for AS and CS (Figure 1B and C). However, secretion was up to 20 times lower than in the BMDMs from wildtype mice, demonstrating the dependence of IL-1β secretion on the NALP3 inflammasome.

After 24 h, IL-1β levels were detectable in the supernatants from NALP3 knockout cells after treatment with all particles. Effects were significant for NT2, AS and CS.

The role of ROS in inflammasome activation in the BMDMs was then explored by EPR spectroscopy. Results of this analysis are shown in Figure 2. Significant ROS increases were detected in LPS pre-activated BMDMs following treatment with all four TiO_2 NMs as well as with CS. Interestingly, no significant increase was seen with AS. Without LPS pre-stimulation, significantly increased EPR signals were observed only with NT1 and NT2. Apart from the positive control H_2O_2, the strongest ROS response was always observed with NT2 (see Figure 2).

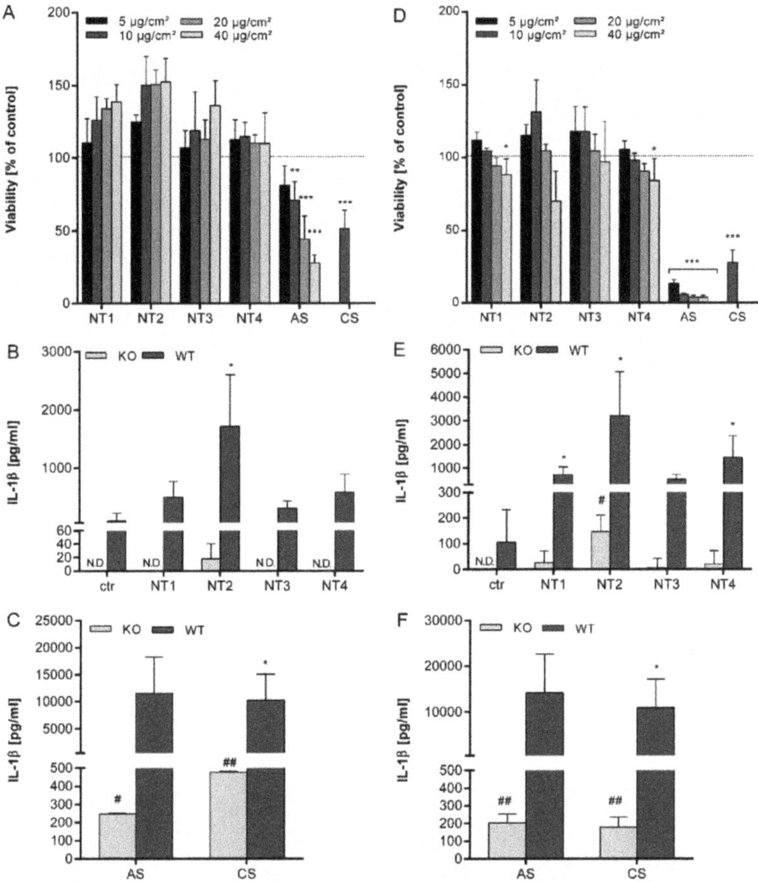

Figure 1. Effects of different TiO_2 NMs on viability and IL-1β secretion after 4 h and 24 h in BMDMs. BMDMs were pre-incubated for 4 h with 10 ng/mL LPS and afterwards treated for 4 h or 24 h with 5–40 µg/cm^2 of four different TiO_2 NMs (NT1, NT2, NT3, NT4) as well as amorphous SiO_2 (AS) or crystalline silica (CS). Viability was detected using WST-1 assay and calculated as percent of control after treatment for 4 h (**A**) or 24 h (**D**). BMDMs of wild type and NALP3 knock out cells were treated for 4 h (**B**,**C**) or 24 h (**E**,**F**) with 40 µg/cm^2 of TiO_2 NMs (**B**,**E**) or amorphous and crystalline silica (**C**,**F**). Release of IL-1β into culture supernatants was detected via ELISA. Mean and standard deviation of three independent experiments are depicted. The asterisks indicate a significant decrease in viability compared to untreated controls or significant increase in IL-1β concentrations compared to wild type controls. (* $p \leq 0.05$; ** $p \leq 0.01$; *** $p \leq 0.001$). The hashes indicate a significant increase of IL-1β concentrations compared to untreated knock-out control cells (# $p \leq 0.05$; ## $p \leq 0.01$).

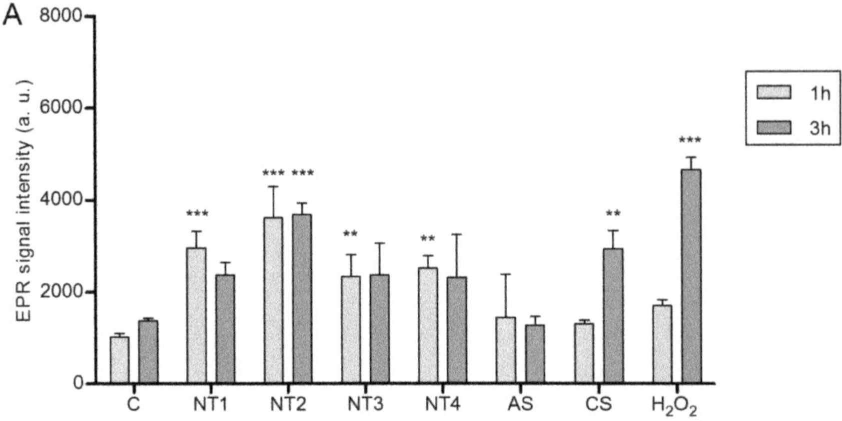

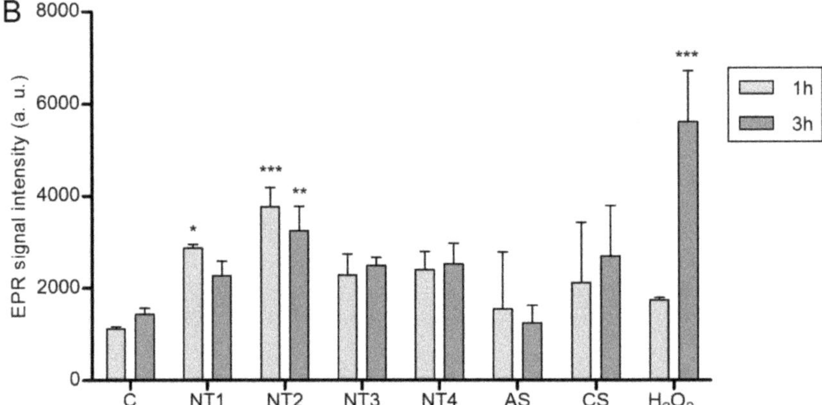

Figure 2. Detection of reactive oxygen species in BMDM via EPR spectroscopy. Cells were either primed for 4 h with 10 ng/mL LPS (**A**) or left unprimed (**B**) prior to treatment with 40 µg/cm^2 of the four TiO_2 NMs, SiO_2 or DQ12 for 1 or 3 h. As a positive control H_2O_2 (6 mM) was used. Simultaneously with these treatments, DMPO was added at a final concentration of 0.1 M. Mean and standard deviation of three independent experiments are depicted. The asterisks indicate a significant change in EPR signal intensity compared to untreated control (* $p \leq 0.05$; ** $p \leq 0{,}01$; *** $p \leq 0.001$).

3.2. *Effects of TiO_2 NMs on NR8383 Rat Alveolar Macrophages*

For the further evaluation of the inflammatory properties of the TiO_2 NMs, experiments were performed with NR8383 rat alveolar macrophages. To evaluate potential direct effects of the NMs on IL-1β mRNA expression and protein release, experiments were performed in the absence of LPS priming. Results are shown in Figure 3. In anticipation of the higher robustness of these immortalized cells compared to the primary BMDMs, particle concentrations from 10 to 80 µg/cm^2 were selected for viability analysis by WST-1 assay. A significant reduction of viability was observed following 24 h treatment at 80 µg/cm^2 for NT1 and NT4 (see Figure 3C). Although cytotoxicity levels are lower in the NR8383 cells, differences between the four TiO_2 are comparable to the results observed with the BMDMs. Accordingly, we choose a treatment concentration of 40 µg/cm^2 for further experiments with the NR8383 cells. As shown in Figure 3A, none of the TiO_2 NMs caused a significant increase in IL-1β mRNA after 4 h of treatment. After 24 h treatment, mRNA levels were significantly increased for NT1 and NT4. Increases in mRNA expression were also noted for NT2 at 24 h, and for AS at both 4 h and 24 h, although these effects did not reach statistical significance. Effects of the particles on IL-1β release from the NR8383 cells are shown in Figure 3B. In the absence of LPS priming, a significant increase in IL-1β release was only observed for AS. IL-1β levels were not significantly increased by the TiO_2 NMs. Subsequent analysis of tumor necrosis factor (TNF) levels in the supernatants revealed findings that were in concordance with the IL-1β protein data. Results are shown in Figure S2 (Supplementary Materials). A significant increase of TNF was observed for AS, but not for TiO_2 NMs.

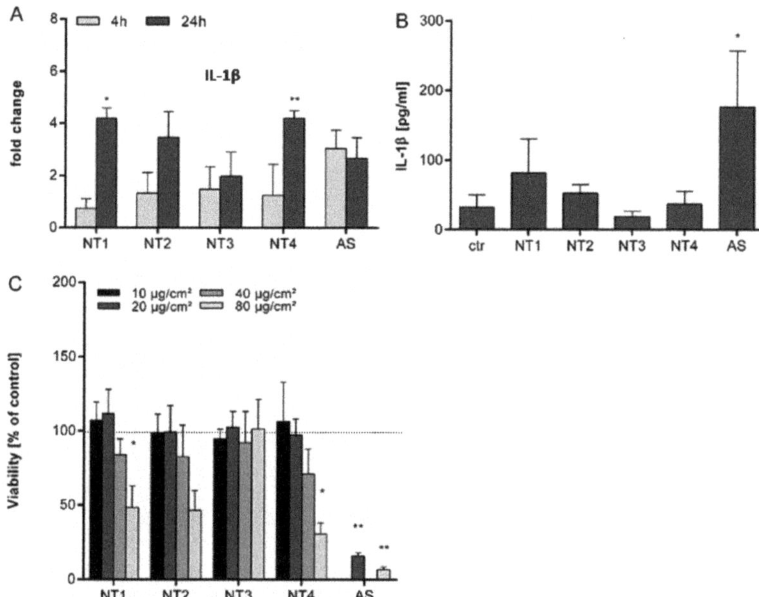

Figure 3. Effects of different TiO_2 NMs on viability, IL-1β mRNA expression and release of IL-1β from NR8383 cells. NR8383 rat alveolar macrophages were treated for 4 h or 24 h with 40 μg/cm^2 of the four different TiO_2 NMs (NT1-4) or amorphous silica (AS). Afterwards, fold changes in IL-1β mRNA expression were analyzed by qRT-PCR in relation to control cells (**A**). For detection of IL-1β in culture supernatants, cells were treated for 24 h and concentrations of IL-1β were detected by ELISA (**B**). Cell viability was evaluated by WST-1 assay following 24 h exposure to TiO_2 NMs (NT1-4) at 10, 20, 40 or 80 μg/cm^2 or AS at 20 or 80 μg/cm^2 (**C**). Mean and standard deviation of three independent experiments are depicted. The asterisks indicate a significant difference in gene expression (**A**), IL-1β secretion (**B**) or viability (**C**) compared to untreated controls. (* $p \leq 0.05$; ** $p \leq 0.01$).

To further explore the observed variability in the inflammatory properties of the TiO_2 NMs, the mRNA expression of HO-1, iNOS and IL-6 were evaluated (Figure 4). For the oxidative stress marker gene HO-1, the most pronounced upregulations were observed with NT2 and amorphous silica after 4 h treatment. At this time point, there were substantial differences in HO-1 mRNA expression between the four TiO_2 NMs (see Figure 4A). After 24 h, the mRNA levels of HO-1 tended to decline to control levels for all used particles. In contrast, the mRNA expression of iNOS increased with increased treatment time. At 24 h, at least 10-fold increased levels were found for all NMs except NT3 (see Figure 4B). Analysis of the mRNA expression of the inflammatory cytokine IL-6 revealed a marked upregulation at 4 h for AS. This effect seemed to be transient as indicated from the lower mRNA levels observed at 24 h for this sample. No such marked effects on IL-6 mRNA expression were observed for the TiO_2 NMs.

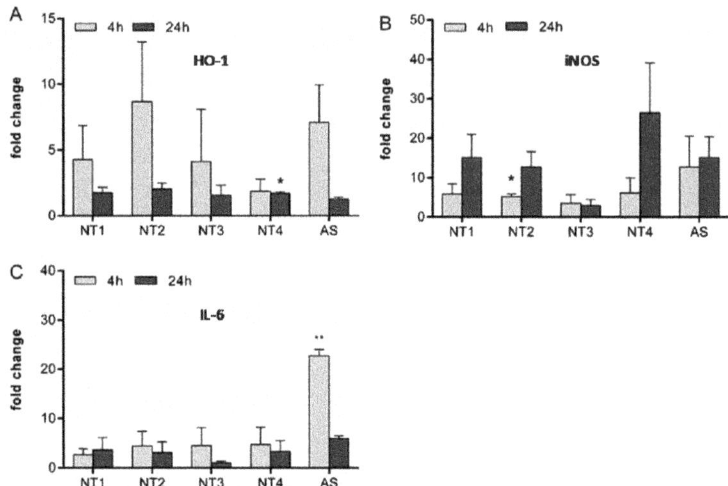

Figure 4. Effects of different TiO_2 NMs on mRNA expression of HO-1, iNOS and IL-6 in NR8383 cells. For analysis of changes in the mRNA regulation of HO-1 (**A**), iNOS (**B**) and IL-6 (**C**), NR8383 cells were treated for 4 h or 24 h with 40 μg/cm^2 of the four TiO_2 NMs (NT1-4) or amorphous silica (AS). Afterwards, mRNA levels were detected via qRT-PCR. Mean and standard deviation of three independent experiments are depicted. The asterisks indicate a significant difference in gene expression compared to untreated controls. (* $p \leq 0.05$; ** $p \leq 0.01$).

3.3. *Evaluation of the Inflammasome Activating Capacity of a Panel of 19 TiO_2 NMs*

To determine the effects of TiO_2 with different characteristics on the secretion of IL-1β, LPS primed NR8383 cells were treated with 40 μg/cm^2 of a large panel of different TiO_2 NMs for 24 h (Figure 5). Compared to control cells, IL-1β concentrations in particle treated cells without priming were not significantly increased. Cells pre-treated with 100 ng/mL LPS without subsequent nanoparticle treatment secreted an increased amount of IL-1β. The additional treatment with the different TiO_2 NMs led to contrasting results. While treatment with some of the used NMs (i.e., NT2, NT13, NT14 and NT17) led to significantly increased IL-1β concentrations in the cell culture supernatant, others did not show comparable effects. In fact, for five out of the nineteen TiO_2 NMs, IL-1β concentrations in the supernatant of cells were not elevated at all (i.e., NT3, NT6, NT9, NT12, NT18). Also, the fine TiO_2 particles that were included in these experiments failed to induce any IL-1β release. In contrast, AS as well as CS showed highly significant increases in IL-1β secretion into the supernatant of the LPS-primed NR8383 cells. Taken together, these data confirmed a considerable diversity of IL-1β secretion for the different TiO_2 NMs used in this study.

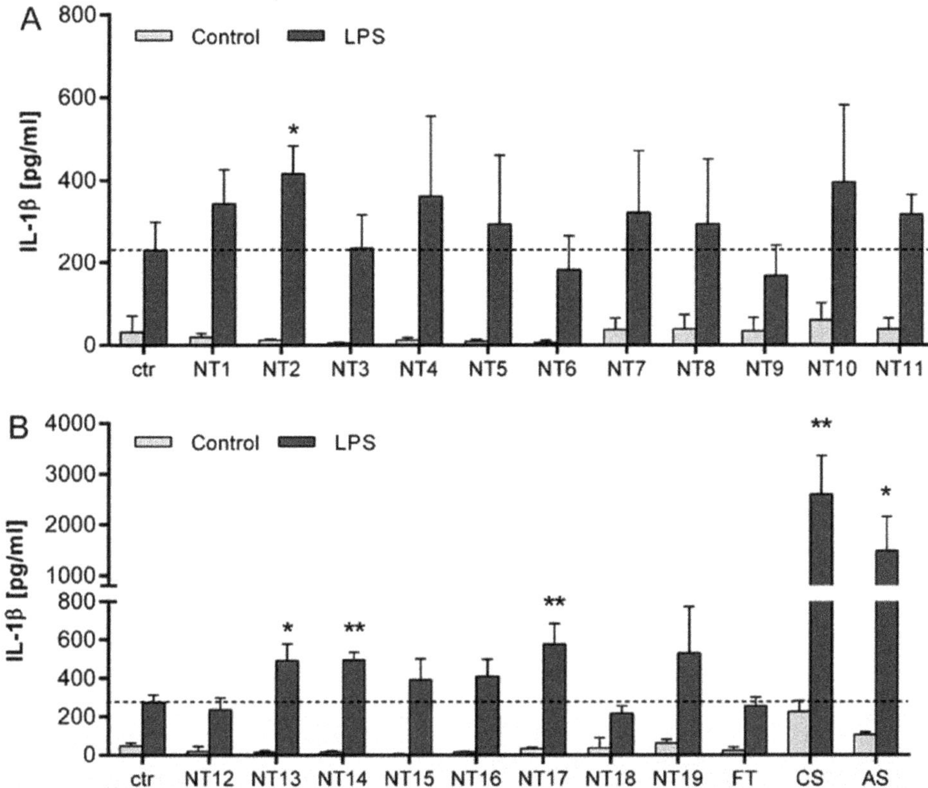

Figure 5. IL-1β release from NR8383 cells following 24 h treatment with TiO_2 NMs. Cells were either primed for 4 h with 100 ng/mL LPS or left without priming, followed by 24 h treatment with 40 µg/cm^2 of different particles. (**A**) IL-1β levels in NR8383 cells treated with TiO_2 NMs (NT1–NT11) from the SetNanoMetro project. (**B**) IL-1β levels in NR8383 cells treated with the TiO_2 NMs from the projects NanOxiMet (NT12- NT15), ENPRA (N16-N19), fine TiO_2 (FT), amorphous SiO_2 (AS) and the crystalline silica sample DQ12 (CS). IL-1β concentrations were analyzed by ELISA. Mean and standard deviation of three independent experiments are depicted. The dashed lines represent the IL-1β concentrations released from control cells. For statistical analysis, the mean values of unprimed cells were subtracted from LPS primed mean values. The asterisks indicate a significant change in IL-1β concentration compared to untreated control (* $p \leq 0.05$; ** $p \leq 0.01$).

4. Discussion

There has been substantial controversy and debate on the toxic and pro-inflammatory properties of TiO_2 NMs, specifically regarding their ability to activate the NALP3 inflammasome. Here, we evaluated the inflammasome-activating properties of a panel of 19 different TiO_2 NMs alongside with a fine crystalline silica and a synthetic amorphous silica sample. The release of IL-1β, as product of NALP3 inflammasome activation was analyzed from mouse BMDMs as well as NR8383 rat alveolar macrophages with or without LPS pre-activation. The BMDMs were included in this study for two reasons. First, the available literature indicates that primary phagocytes, like BMDMs or BMDCs, are more sensitive to TiO_2 NMs than immortalized cell lines regarding IL-1β release [20,21,23]. Moreover, the use of BMDMs allowed us to directly compare the IL-1β releasing properties of the NMs in primary cells from mice with or without functional NALP3 inflammasome.

In the NALP3 proficient BMDMs, increased IL-1β release was observed for TiO_2 NMs upon LPS pre-treatment, albeit at different levels among the four investigated samples. First, our findings demonstrate that nanoparticulate TiO_2 is able to induce the release of IL-1β from macrophages at sub-cytotoxic concentrations, as verified by WST-1 assay. Second, the contrasting levels of IL-1β release suggest that the ability of TiO_2 NMs to trigger NALP3 inflammasome activation depends on their particle characteristics. Comparable results were observed by other investigators [8], using the human macrophage-like cell line THP-1. In our hands, effects of the amorphous silica and the crystalline silica on IL-1β release were much stronger than the effects of the TiO_2 NMs at equal mass dose. This is in support of the strong difference in the inflammatory potency between these respective particle types. Considering the surface area as dose metric, the IL-1β releasing potency of the crystalline silica is markedly higher than that of the amorphous silica while that of the TiO_2 NMs was lower than that of the amorphous SiO_2. A strong difference in surface area-adjusted inflammatory potency between crystalline silica and TiO_2 NMs was previously also found for interleukin-8 release from lung epithelial cells [10]. For the four TiO_2 NMs, the IL-1β secretion after 24 h treatment was higher compared to 4 h treatment, revealing a time dependent effect of this pro-inflammatory response. However, levels of IL-1β at 24 h were only twice as high as after 4 h treatment. Also for amorphous and crystalline silica the levels of IL-1β did not differ substantially between 4 h and 24 h. These findings are in alignment with the role of IL-1β as an early mediator in inflammation [36].

The NALP3 inflammasome dependence of the IL-1β secretion by the TiO_2 NMs was demonstrated by our findings with the BMDMs of NALP3 deficient mice. Compared to the BMDMs from the NALP3 proficient mice a massive reduction of IL-1β release was observed with the deficient macrophages. Our findings also point towards an alternative, NALP3 independent mechanism of IL-1β processing, as indicated by the time-dependent IL-1β release from knockout BMDMs upon treatment with NT2 (as well with the amorphous and crystalline silica). This mechanism possibly involves a direct processing of pro IL-1β by cathepsins [37]. The experiments with NALP3 deficient macrophages also confirmed the well-established inflammasome activating capacity of crystalline silica [13,17,19] as well of amorphous silica [38–40]. Complementary investigations with a commercial luminescence-based casapase-1 activity kit provided further support for the contrasting inflammasome activating properties of these different particle types. In supernatants of NALP3-proficient BMDMs we found no significant increases in caspase-1 activity upon treatment with the TiO_2 NMs in contrast to the crystalline silica, while an intermediate effect was observed for the amorphous silica sample (data not shown). Unfortunately, we could not reliably analyze caspase-1 activity within BMDM cell lysates, likely as a results of assay interference with endocytosed particles. Further research is needed to elaborate on the mechanisms of NALP (in)dependent IL-1β maturation in particle exposed BMDMs.

Generation of ROS has been linked to the inflammatory properties of NMs, including their ability to activate the NALP3 inflammasome [16–18,40]. Therefore, we also determined ROS levels in the particle treated BMDMs by EPR. Indeed, increased ROS levels were observed following treatment with the TiO_2 NMs. Moreover, the strongest effect was observed for NT2, the sample that also showed the strongest NALP3-dependent IL-1β release, in support of the role of ROS in NALP3 inflammation by nanoparticulate TiO_2. The involvement of ROS was further substantiated by complementary investigations, revealing a significant inhibition of IL-1β release in NT2 treated BMDMs that were pre-treated with the antioxidant diphenyleneiodonium (DPI) (data not shown).

Notably, in the EPR analyses the ROS increases observed with the TiO_2 NMs did not substantially differ from those observed with AS and CS, even though the latter two particles had a much more pronounced IL-1β response. This indicates that TiO_2 NMs, AS and CS, at least in part, elicit NALP inflammation through distinct mechanisms. The marked differences in cytotoxicity between these types of particles and their impact on endogenous ROS sources should be taken into account here.

Our subsequent investigations were performed with NR8383 cells, a well-established alveolar macrophage cell line from rat [41]. NR8383 cells have emerged as a reliable in vitro model to study the mechanisms of toxicity of various particles including crystalline and amorphous silica as well as

TiO_2 particles [25,42,43]. Obviously, in vitro systems never fully recapitulate the in vivo situation but nevertheless can be very useful for hazard ranking. Indeed, the NR8383 cell line has also proven to be useful for (nano)particle hazard grouping strategies, showing good comparability with outcomes from in vivo studies [44–46]. To evaluate the suitability of the NR8383 cells for the investigation of the inflammasome activating properties of the TiO_2 NMs, at first, experiments were performed in the absence of LPS priming. In this case, no increased IL-1β concentrations were found following treatment with any of the four TiO_2 NMs. In contrast, treatment with amorphous SiO_2 led to a significant increase in IL-1β secretion from the non-LPS-primed NR8383 cells. In concordance with this, also no significant release of the early inflammatory cytokine TNF was found with the TiO_2 NMs, again in contrast to the amorphous SiO_2 sample (Figure S2). Crystalline silica was not included in these NR8383 experiments, but its ability to induce IL-1β release (as well as TNF) in the absence of LPS-priming has been well-documented for this cell line [25,47]. Taken together, our data suggest that TiO_2 NMs are incapable of inducing a full immunological activation without LPS priming. Activation of the IL1B gene by particles is mediated by the activation of the transcription factors nuclear factor-κB (NFκB) and activator protein 1 (AP-1) [17]. Accordingly, the mRNA expression analysis of the non-LPS-primed NR8383 cells further supported the poor activating properties of the TiO_2 NMs. At early time point (4 h), elevated IL-1β mRNA levels were only observed with the amorphous SiO_2. At 24 h, statistically significant increases in IL-1β mRNA expression were observed with NT1 and NT4 but overall the effects in the NR8383 cells were modest, in alignment with the poor ability of TiO_2 to activate NFκB in these cells, shown previously [25].

To further analyze the effects of TiO_2 NMs on oxidative stress induction and inflammation in the NR83838 cells, the mRNA expression levels of HO-1, iNOS and IL-6 were analyzed. HO-1 is considered as a sensitive marker of oxidative stress, and its mRNA expression has been found upregulated in lungs of crystalline-silica exposed rats [47] as well as mice [48]. In our present study, HO-1 mRNA levels were enhanced following 4 h treatment with the amorphous SiO_2 and all TiO_2 NMs, except NT4, indicative of oxidative stress by these NMs. The absence of marked increases in HO-1 mRNA expression levels at the 24 h time point likely relates to a compensatory upregulation of antioxidant defense responses [49,50]. In contrast, the mRNA levels of iNOS were most strongly increased after 24 h treatment for all NMs, with the exception of NT3. Considering the transcriptional activation of this gene as a marker of inflammation, as revealed for crystalline silica [47,51], current iNOS mRNA data fit well with the observed release of IL-1β. Both in the NR8383 cells and the BMDMs, NT3 caused the lowest IL-1β secretion among the four TiO_2 NMs. Our current data also align with previous findings in our lab, showing increased mRNA expression of HO-1 as well as iNOS in NR8383 cells treated with crystalline silica and TiO_2 NMs, while fine TiO_2 particles were not reactive [25]. In our current study, unlike HO-1 and iNOS, the mRNA expression analysis of IL-6 did not reveal marked contrasts between the different TiO_2 NP. However, a substantial up-regulation was observed in the NR8383 cells treated with the amorphous silica. As such, these data align well with the secretion pattern of IL-1β (and TNFα), and further supports the contrasting pro-inflammatory potency of this nano-SiO_2, when compared to the nano-TiO_2.

The importance of the physicochemical properties of particles on inflammasome activation has been reviewed recently [17] and shown for TiO_2 in THP-1 macrophages [8]. In our hands, neither the characteristics of the pristine samples (Table 1), nor their morphological behavior in suspensions (Table 2), provided any plausible explanation for the observed differences in the IL-1β generating potency of the TiO_2 NMs. The two samples that showed the highest contrast in IL-1β release, i.e., NT2 and NT3, both were pure anatase phase and were also nearly identical in primary particle diameter as well as their specific particle surface area (BET). Furthermore, there was no apparent correlation between IL-1β release and the agglomeration states of the 4 samples. Here the strong differences in agglomeration state for the different culture media used for the BMDMs and the NR8383 should also be highlighted. As a result of the differences in media composition, the four TiO_2 samples ranked differently regarding agglomerate size. Yet, for both cell types, NT2 turned out to be the most potent in terms of inflammasome activation and IL-1β release. The differences in agglomeration status of the TiO_2 NMs in the different culture media can be explained by the absence of the FCS in the "DMEM" used for the NR8383 cells, and aligns with previous investigations by Allouni and coworkers [52].

This observation is all the more interesting, as it suggests that some TiO_2 NMs can trigger IL-1β release from LPS primed macrophages, irrespective of the presence of an abundant amount of proteins in the cell culture medium. It has been well-established that the in vitro effects of NMs, including TiO_2 and SiO_2 may strongly depend on the presence of FCS [31,53–55]. Therefore, further research is warranted to determine the role of corona-forming constituents in the inflammasome pathway activating properties of inhaled TiO_2 NMs, or other types of particles like the amorphous and crystalline silica that were used as positive control in our study. In this regard it will be particularly relevant to evaluate the effects of realistic lung lining fluids, such as the artificial bronchoalveolar lavage fluid (BALF) introduced by Porter and coworkers [56].

An independent property that possibly drives inflammasome activating properties of TiO_2 is shape. Indeed, in a side-by-side toxicological evaluation of long versus shortened TiO_2 nanofibers, Bianchi and colleagues recently showed an increased pro-inflammatory response in mouse lung and peritoneum for the longer samples [57]. Earlier, Porter et al. [58] evaluated the role of TiO_2 NM shape and length upon pharyngeal application in mouse lung by comparison of the effects of TiO_2 nanospheres and short and long nanobelts with respective aspect ratios of 1:30 and 1:80. They identified persistent inflammatory effects for the nanobelts, with the longer one being more potent and also exclusively causing significant lung tissue remodeling (i.e., fibrosis). While these studies demonstrated the importance of length in the pro-inflammatory potency of so-called high aspect ratio nanomaterials (HARN) composed of TiO_2, they cannot provide a direct explanation for our current study findings. The TiO_2 NMs samples in our study were composed of primary particles with all three external dimensions in the nanoscale and do not categorize as HARN (data not shown).

In contrast to the aforementioned physicochemical properties of the TiO_2 NMs, their ability to induce ROS generation and associated oxidative stress was found to be a good predictor of the inflammasome activating properties in our hands. Both in the BMDMs and the NR8383 cells, the strongest IL-1β release in LPS-primed conditions was observed with the sample NT2. This sample also displayed the highest and most sustained ROS levels during treatment of the BMDMs, as well as the highest mRNA expression levels of the oxidative stress marker gene HO-1 in the NR8383 cells.

Finally, to prove the feasibility of NR8383 cells for the hazard screening of NMs, specifically regarding their inflammasome activating capacity, we selected a large panel of particles. In addition to the four in-depth-investigated TiO_2 NMs, the amorphous silica and the crystalline silica sample DQ12, 15 further TiO_2 NMs were included as well as a sample of fine TiO_2. For the six samples that were studied in both in vitro models a remarkable concordance was found in terms of qualitative responses: The TiO_2 sample NT3 did not cause an increased IL-1β release upon LPS priming in both cell models whereas NT2, and more importantly, the positive controls CS and AS did. Again, one should keep in mind here that the NR8383 cells were treated in the absence of FCS to inhibit undesired proliferation of this robust cell line, whereas the primary mouse BMDMs required treatment in FCS containing medium. Thus, while the effects appeared qualitatively similar, the contrasting abundance of corona forming proteins in the respective treatment media may very well explain for some of the observed quantitative differences in IL1β responses between both in vitro models. Moreover, differences in the constitutive or LPS-induced release of IL1β between cell lines and primary cells or between murine and human cells may also play a role.

The outcome of the NR8383 experiments with the large panel of particles further confirmed our observations about the variation in responsiveness of macrophages to TiO_2 NMs. Indeed, the available literature already points to a heterogeneity in effects of TiO_2 NMs. However, for our present findings we can rule out any potential effects that could result from differences in the selected macrophage cell type and its culturing protocol as well as the particle handling procedure and administration protocol. Among the 19 tested nano-TiO_2 samples, four samples showed a significant increase in IL-1β release from the LPS-primed NR8383 macrophages while five nano-TiO_2 samples did not show any increase above control. The remaining ten TiO_2 NMs could be arbitrarily grouped as "intermediate". Again, it should be emphasized that, while displaying a clear gradient in IL-1β releasing potency, the responsiveness of the NR8383 cells to amorphous and crystalline silica was much greater. Moreover, when considered per unit surface area, the fine crystalline silica was much more potent than the synthetic amorphous silica. Taken together, these data suggest that evaluation of the inflammasome generating properties in the immortalized NR8383 rat alveolar macrophage cell line is a useful approach for a pro-inflammatory hazard ranking of inhaled nanomaterials. This obviously includes the assessment of NMs of different classes and chemical composition, but likely also even within an individual group of NMs of the same chemical composition, as exemplified here by the case of TiO_2.

5. Conclusions

Our study demonstrates that specific TiO_2 nanomaterials can activate the inflammasome in macrophages in association with ROS generation, albeit with a markedly lower potency than the synthetic amorphous silica and the crystalline silica. The heterogeneity in IL-1β release among 19 different TiO_2 samples observed in our study underscores the need for case-by-case evaluation of the inflammasome activating capacity of NMs. Our data also suggests that the NR8383 cell line can serve as a robust in vitro tool for such evaluation.

Supplementary Materials: The following are available online at http://www.mdpi.com/2079-4991/10/9/1876/s1, Figure S1: IL-1β concentration in cell culture supernatants of BMDMs after 24 h treatment with different TiO_2 NMs, amorphous silica and crystalline silica. Figure S2: Tumor Necrosis Factor concentrations in supernatants of NR8383 cell cultures after 24 h treatment with different TiO_2 NMs and amorphous silica.

Author Contributions: J.K. performed all NR8383 cell culture experiments and qRT-PCR analyses, J.T. performed all experiments with the BMDMs and EPR analyses, J.K. and J.T. performed WST-1 assays and ELISA measurements, B.H. performed the DLS analysis and provided advice regarding the EPR experiments, J.K., J.T. and R.P.F.S. wrote and reviewed the manuscript, C.A. participated in the design of experiments and study coordination, R.P.F.S. devised the project, coordinated and supervised the study. All authors have read and agreed to the published version of the manuscript.

Funding: The work leading to these results has received funding from the European Union's Seventh Framework Programme for research, technology development and demonstration under grant agreement n° 604577 (SETNanoMetro, FP7-NMP-2013-LARGE-7).

Acknowledgments: Specific types of TiO_2 NMs were kindly provided by Solaronix (Switzerland), Evonik, Cristal and the University of Turin (Italy) in the framework of the EU-FP7 project SETNanoMetro (FP7-NMP-2013-LARGE-7). The authors thank Christel Weishaupt and Gaby Wick for technical support.

Conflicts of Interest: The authors declare no conflict of interest. The funders had no role in the design of the study; in the collection, analyses, or interpretation of data; in the writing of the manuscript, or in the decision to publish the results.

References

1. Robichaud, C.O.; Uyar, A.E.; Darby, M.R.; Zucker, L.G.; Wiesner, M.R. Estimates of upper bounds and trends in nano-TiO_2 production as a basis for exposure assessment. *Environ. Sci. Technol.* **2009**, *43*, 4227–4233. [CrossRef] [PubMed]
2. Winkler, H.C.; Notter, T.; Meyer, U.; Naegeli, H. Critical review of the safety assessment of titanium dioxide additives in food. *J. Nanobiotechnol.* **2018**, *16*, 51. [CrossRef] [PubMed]
3. Donaldson, K.E.N.; Borm, P.J.A. The Quartz Hazard: A Variable Entity. *Ann. Occup. Hyg.* **1998**, *42*, 287–294. [CrossRef]
4. Albrecht, C.; Schins, R.P.F.; Höhr, D.; Becker, A.; Shi, T.; Knaapen, A.M.; Borm, P.J.A. Inflammatory time course after quartz instillation: Role of tumor necrosis factor-alpha and particle surface. *Am. J. Respir. Cell Mol. Biol.* **2004**, *31*, 292–301. [CrossRef]
5. Fubini, B. Surface Chemistry and Quartz Hazard. *Ann. Occup. Hyg.* **1998**, *42*, 521–530. [CrossRef]

6. Freyre-Fonseca, V.; Delgado-Buenrostro, N.L.; Gutiérrez-Cirlos, E.B.; Calderón-Torres, C.M.; Cabellos-Avelar, T.; Sánchez-Pérez, Y.; Pinzón, E.; Torres, I.; Molina-Jijón, E.; Zazueta, C.; et al. Titanium dioxide nanoparticles impair lung mitochondrial function. *Toxicol. Lett.* **2011**, *202*, 111–119. [CrossRef]
7. Grassian, V.H.; O'shaughnessy, P.T.; Adamcakova-Dodd, A.; Pettibone, J.M.; Thorne, P.S. Inhalation exposure study of titanium dioxide nanoparticles with a primary particle size of 2 to 5 nm. *Environ. Health Perspect.* **2007**, *115*, 397–402. [CrossRef]
8. Morishige, T.; Yoshioka, Y.; Tanabe, A.; Yao, X.; Tsunoda, S.-I.; Tsutsumi, Y.; Mukai, Y.; Okada, N.; Nakagawa, S. Titanium dioxide induces different levels of IL-1beta production dependent on its particle characteristics through caspase-1 activation mediated by reactive oxygen species and cathepsin B. *Biochem. Biophys. Res. Commun.* **2010**, *392*, 160–165. [CrossRef]
9. Oberdörster, G.; Ferin, J.; Lehnert, B.E. Correlation between particle size, in vivo particle persistence, and lung injury. *Environ. Health Perspect.* **1994**, *102* (Suppl. S5), 173–179.
10. Singh, S.; Shi, T.; Duffin, R.; Albrecht, C.; van Berlo, D.; Höhr, D.; Fubini, B.; Martra, G.; Fenoglio, I.; Borm, P.J.A.; et al. Endocytosis, oxidative stress and IL-8 expression in human lung epithelial cells upon treatment with fine and ultrafine TiO2: Role of the specific surface area and of surface methylation of the particles. *Toxicol. Appl. Pharm.* **2007**, *222*, 141–151. [CrossRef]
11. Nurkiewicz, T.R.; Porter, D.W.; Hubbs, A.F.; Cumpston, J.L.; Chen, B.T.; Frazer, D.G.; Castranova, V. Nanoparticle inhalation augments particle-dependent systemic microvascular dysfunction. *Part. Fibre Toxicol.* **2008**, *5*, 1. [CrossRef] [PubMed]
12. Cassel, S.L.; Eisenbarth, S.C.; Iyer, S.S.; Sadler, J.J.; Colegio, O.R.; Tephly, L.A.; Carter, A.B.; Rothman, P.B.; Flavell, R.A.; Sutterwala, F.S. The Nalp3 inflammasome is essential for the development of silicosis. *Proc. Natl. Acad. Sci. USA* **2008**, *105*, 9035–9040. [CrossRef] [PubMed]
13. Dostert, C.; Pétrilli, V.; van Bruggen, R.; Steele, C.; Mossman, B.T.; Tschopp, J. Innate immune activation through Nalp3 inflammasome sensing of asbestos and silica. *Science* **2008**, *320*, 674–677. [CrossRef] [PubMed]
14. Hornung, V.; Bauernfeind, F.; Halle, A.; Samstad, E.O.; Kono, H.; Rock, K.L.; Fitzgerald, K.A.; Latz, E. Silica crystals and aluminum salts activate the NALP3 inflammasome through phagosomal destabilization. *Nat. Immunol.* **2008**, *9*, 847–856. [CrossRef] [PubMed]
15. Martinon, F.; Burns, K.; Tschopp, J. The Inflammasome. *Mol. Cell* **2002**, *10*, 417–426. [CrossRef]
16. Hamilton, R.F.; Thakur, S.A.; Holian, A. Silica binding and toxicity in alveolar macrophages. *Free. Radic. Biol. Med.* **2008**, *44*, 1246–1258. [CrossRef] [PubMed]
17. Rabolli, V.; Lison, D.; Huaux, F. The complex cascade of cellular events governing inflammasome activation and IL-1β processing in response to inhaled particles. *Part. Fibre Toxicol.* **2016**, *13*, 40. [CrossRef] [PubMed]
18. Sohaebuddin, S.K.; Thevenot, P.T.; Baker, D.; Eaton, J.W.; Tang, L. Nanomaterial cytotoxicity is composition, size, and cell type dependent. *Part. Fibre Toxicol.* **2010**, *7*, 22. [CrossRef] [PubMed]
19. Peeters, P.M.; Eurlings, I.M.J.; Perkins, T.N.; Wouters, E.F.; Schins, R.P.F.; Borm, P.J.A.; Drommer, W.; Reynaert, N.L.; Albrecht, C. Silica-induced NLRP3 inflammasome activation in vitro and in rat lungs. *Part. Fibre Toxicol.* **2014**, *11*, 58. [CrossRef]
20. Baron, L.; Gombault, A.; Fanny, M.; Villeret, B.; Savigny, F.; Guillou, N.; Panek, C.; Le Bert, M.; Lagente, V.; Rassendren, F.; et al. The NLRP3 inflammasome is activated by nanoparticles through ATP, ADP and adenosine. *Cell Death Dis.* **2015**, *6*, e1629. [CrossRef]
21. Riedle, S.; Pele, L.C.; Otter, D.E.; Hewitt, R.E.; Singh, H.; Roy, N.C.; Powell, J.J. Pro-inflammatory adjuvant properties of pigment-grade titanium dioxide particles are augmented by a genotype that potentiates interleukin 1β processing. *Part. Fibre Toxicol.* **2017**, *14*, 51. [CrossRef] [PubMed]
22. Ruiz, P.A.; Morón, B.; Becker, H.M.; Lang, S.; Atrott, K.; Spalinger, M.R.; Scharl, M.; Wojtal, K.A.; Fischbeck-Terhalle, A.; Frey-Wagner, I.; et al. Titanium dioxide nanoparticles exacerbate DSS-induced colitis: Role of the NLRP3 inflammasome. *Gut* **2017**, *66*, 1216–1224. [CrossRef] [PubMed]
23. Winter, M.; Beer, H.-D.; Hornung, V.; Krämer, U.; Schins, R.P.F.; Förster, I. Activation of the inflammasome by amorphous silica and TiO_2 nanoparticles in murine dendritic cells. *Nanotoxicology* **2011**, *5*, 326–340. [CrossRef] [PubMed]
24. Reisetter, A.C.; Stebounova, L.V.; Baltrusaitis, J.; Powers, L.; Gupta, A.; Grassian, V.H.; Monick, M.M. Induction of inflammasome-dependent pyroptosis by carbon black nanoparticles. *J. Biol. Chem.* **2011**, *286*, 21844–21852. [CrossRef]

25. Scherbart, A.M.; Langer, J.; Bushmelev, A.; van Berlo, D.; Haberzettl, P.; van Schooten, F.-J.; Schmidt, A.M.; Rose, C.R.; Schins, R.P.F.; Albrecht, C. Contrasting macrophage activation by fine and ultrafine titanium dioxide particles is associated with different uptake mechanisms. *Part. Fibre Toxicol.* **2011**, *8*, 31. [CrossRef] [PubMed]
26. Tsugita, M.; Morimoto, N.; Nakayama, M. SiO_2 and TiO_2 nanoparticles synergistically trigger macrophage inflammatory responses. *Part. Fibre Toxicol.* **2017**, *14*, 11. [CrossRef]
27. Marucco, A.; Carella, E.; Fenoglio, I. A comparative study on the efficacy of different probes to predict the photo-activity of nano-titanium dioxide toward biomolecules. *RSC Adv.* **2015**, *5*, 89559–89568. [CrossRef]
28. Deiana, C.; Minella, M.; Tabacchi, G.; Maurino, V.; Fois, E.; Martra, G. Shape-controlled TiO_2nanoparticles and TiO_2 P_{25} interacting with CO and H_2O_2 molecular probes: A synergic approach for surface structure recognition and physico-chemical understanding. *Phys. Chem. Chem. Phys.* **2013**, *15*, 307. [CrossRef]
29. Iannarelli, L.; Giovannozzi, A.M.; Morelli, F.; Viscotti, F.; Bigini, P.; Maurino, V.; Spoto, G.; Martra, G.; Ortel, E.; Hodoroaba, V.D.; et al. Shape engineered TiO_2 nanoparticles in Caenorhabditis elegans: A Raman imaging based approach to assist tissue-specific toxicological studies. *RSC Adv.* **2016**, *6*, 70501–70509. [CrossRef]
30. Kermanizadeh, A.; Gosens, I.; MacCalman, L.; Johnston, H.; Danielsen, P.H.; Jacobsen, N.R.; Lenz, A.-G.; Fernandes, T.; Schins, R.P.F.; Cassee, F.R.; et al. A Multilaboratory Toxicological Assessment of a Panel of 10 Engineered Nanomaterials to Human Health—ENPRA Project—The Highlights, Limitations, and Current and Future Challenges. *J. Toxicol. Environ. Health B Crit. Rev.* **2016**, *19*, 1–28. [CrossRef]
31. Thongkam, W.; Gerloff, K.; van Berlo, D.; Albrecht, C.; Schins, R.P.F. Oxidant generation, DNA damage and cytotoxicity by a panel of engineered nanomaterials in three different human epithelial cell lines. *Mutagenesis* **2017**, *32*, 105–115. [CrossRef] [PubMed]
32. Jessop, F.; Hamilton, R.F.; Rhoderick, J.F.; Fletcher, P.; Holian, A. Phagolysosome acidification is required for silica and engineered nanoparticle-induced lysosome membrane permeabilization and resultant NLRP3 inflammasome activity. *Toxicol. Appl. Pharm.* **2017**, *318*, 58–68. [CrossRef] [PubMed]
33. Albrecht, C.; Borm, P.J.A.; Adolf, B.; Timblin, C.R.; Mossman, B.T. In Vitro and in Vivo Activation of Extracellular Signal-Regulated Kinases by Coal Dusts and Quartz Silica. *Toxicol. Appl. Pharm.* **2002**, *184*, 37–45. [CrossRef]
34. Kovarova, M.; Hesker, P.R.; Jania, L.; Nguyen, M.; Snouwaert, J.N.; Xiang, Z.; Lommatzsch, S.E.; Huang, M.T.; Ting, J.P.-Y.; Koller, B.H. NLRP1-dependent pyroptosis leads to acute lung injury and morbidity in mice. *J. Immunol.* **2012**, *189*, 2006–2016. [CrossRef]
35. Livak, K.J.; Schmittgen, T.D. Analysis of relative gene expression data using real-time quantitative PCR and the 2(-Delta Delta C(T)) Method. *Methods* **2001**, *25*, 402–408. [CrossRef]
36. Dinarello, C.A. Immunological and inflammatory functions of the interleukin-1 family. *Annu. Rev. Immunol.* **2009**, *27*, 519–550. [CrossRef]
37. Kono, H.; Orlowski, G.M.; Patel, Z.; Rock, K.L. The IL-1-dependent sterile inflammatory response has a substantial caspase-1-independent component that requires cathepsin C. *J. Immunol.* **2012**, *189*, 3734–3740. [CrossRef]
38. Kusaka, T.; Nakayama, M.; Nakamura, K.; Ishimiya, M.; Furusawa, E.; Ogasawara, K. Effect of silica particle size on macrophage inflammatory responses. *PLoS ONE* **2014**, *9*, e92634. [CrossRef]
39. Rabolli, V.; Badissi, A.A.; Devosse, R.; Uwambayinema, F.; Yakoub, Y.; Palmai-Pallag, M.; Lebrun, A.; de Gussem, V.; Couillin, I.; Ryffel, B.; et al. The alarmin IL-1α is a master cytokine in acute lung inflammation induced by silica micro- and nanoparticles. *Part. Fibre Toxicol.* **2014**, *11*, 69. [CrossRef]
40. Sandberg, W.J.; Låg, M.; Holme, J.A.; Friede, B.; Gualtieri, M.; Kruszewski, M.; Schwarze, P.E.; Skuland, T.; Refsnes, M. Comparison of non-crystalline silica nanoparticles in IL-1β release from macrophages. *Part. Fibre Toxicol.* **2012**, *9*, 32. [CrossRef]
41. Helmke, R.J.; Boyd, R.L.; German, V.F.; Mangos, J.A. From growth factor dependence to growth factor responsiveness: The genesis of an alveolar macrophage cell line. *Vitr. Cell. Dev. Biol.* **1987**, *23*, 567–574. [CrossRef] [PubMed]
42. Ghiazza, M.; Scherbart, A.M.; Fenoglio, I.; Grendene, F.; Turci, F.; Martra, G.; Albrecht, C.; Schins, R.P.F.; Fubini, B. Surface iron inhibits quartz-induced cytotoxic and inflammatory responses in alveolar macrophages. *Chem. Res. Toxicol.* **2011**, *24*, 99–110. [CrossRef] [PubMed]

43. Haberzettl, P.; Duffin, R.; Krämer, U.; Höhr, D.; Schins, R.P.F.; Borm, P.J.A.; Albrecht, C. Actin plays a crucial role in the phagocytosis and biological response to respirable quartz particles in macrophages. *Arch. Toxicol.* **2007**, *81*, 459–470. [CrossRef] [PubMed]
44. Bannuscher, A.; Hellack, B.; Bahl, A.; Laloy, J.; Herman, H.; Stan, M.S.; Dinischiotu, A.; Giusti, A.; Krause, B.-C.; Tentschert, J.; et al. Metabolomics profiling to investigate nanomaterial toxicity in vitro and in vivo. *Nanotoxicology* **2020**, 1–20. [CrossRef]
45. Horie, M.; Tabei, Y.; Sugino, S.; Fukui, H.; Nishioka, A.; Hagiwara, Y.; Sato, K.; Yoneda, T.; Tada, A.; Koyama, T. Comparison of the effects of multiwall carbon nanotubes on the epithelial cells and macrophages. *Nanotoxicology* **2019**, *13*, 861–878. [CrossRef] [PubMed]
46. Wiemann, M.; Vennemann, A.; Sauer, U.G.; Wiench, K.; Ma-Hock, L.; Landsiedel, R. An in vitro alveolar macrophage assay for predicting the short-term inhalation toxicity of nanomaterials. *J. Nanobiotechnol.* **2016**, *14*, 16. [CrossRef]
47. van Berlo, D.; Knaapen, A.M.; van Schooten, F.-J.; Schins, R.P.; Albrecht, C. NF-kappaB dependent and independent mechanisms of quartz-induced proinflammatory activation of lung epithelial cells. *Part. Fibre Toxicol.* **2010**, *7*, 13. [CrossRef]
48. van Berlo, D.; Wessels, A.; Boots, A.W.; Wilhelmi, V.; Scherbart, A.M.; Gerloff, K.; van Schooten, F.J.; Albrecht, C.; Schins, R.P.F. Neutrophil-derived ROS contribute to oxidative DNA damage induction by quartz particles. *Free. Radic. Biol. Med.* **2010**, *49*, 1685–1693. [CrossRef]
49. Rothfuss, A.; Speit, G. Overexpression of heme oxygenase-1 (HO-1) in V79 cells results in increased resistance to hyperbaric oxygen (HBO)-induced DNA damage. *Environ. Mol. Mutagen.* **2002**, *40*, 258–265. [CrossRef]
50. Virág, L.; Jaén, R.I.; Regdon, Z.; Boscá, L.; Prieto, P. Self-defence of macrophages against oxidative injury: Fighting for their own survival. *Redox Biol.* **2019**, *26*, 101261. [CrossRef]
51. Porter, D.W.; Millecchia, L.; Robinson, V.A.; Hubbs, A.; Willard, P.; Pack, D.; Ramsey, D.; McLaurin, J.; Khan, A.; Landsittel, D.; et al. Enhanced nitric oxide and reactive oxygen species production and damage after inhalation of silica. *Am. J. Physiol. Lung Cell. Mol. Physiol.* **2002**, *283*, L485–L493. [CrossRef] [PubMed]
52. Allouni, Z.E.; Cimpan, M.R.; Høl, P.J.; Skodvin, T.; Gjerdet, N.R. Agglomeration and sedimentation of TiO_2 nanoparticles in cell culture medium. *Colloids Surf. B Biointerfaces* **2009**, *68*, 83–87. [CrossRef] [PubMed]
53. Panas, A.; Marquardt, C.; Nalcaci, O.; Bockhorn, H.; Baumann, W.; Paur, H.R.; Mülhopt, S.; Diabaté, S.; Weiss, C. Screening of different metal oxide nanoparticles reveals selective toxicity and inflammatory potential of silica nanoparticles in lung epithelial cells and macrophages. *Nanotoxicology* **2013**, *7*, 259–273. [CrossRef] [PubMed]
54. Vranic, S.; Gosens, I.; Jacobsen, N.R.; Jensen, K.A.; Bokkers, B.; Kermanizadeh, A.; Stone, V.; Baeza-Squiban, A.; Cassee, F.R.; Tran, L.; et al. Impact of serum as a dispersion agent for in vitro and in vivo toxicological assessments of TiO2 nanoparticles. *Arch. Toxicol.* **2017**, *91*, 353–363. [CrossRef]
55. Leibe, R.; Hsiao, I.L.; Fritsch-Decker, S.; Kielmeier, U.; Wagbo, A.M.; Voss, B.; Schmidt, A.; Hessman, S.D.; Duschl, A.; Oostingh, G.J.; et al. The protein corona suppresses the cytotoxic and pro-inflammatory response in lung epithelial cells and macrophages upon exposure to nanosilica. *Arch. Toxicol.* **2019**, *93*, 871–885. [CrossRef]
56. Porter, D.; Sriram, K.; Wolfarth, M.; Jefferson, A.; Schwegler-Berry, D.; Andrew, M.E.; Castranova, V. A biocompatible medium for nanoparticle dispersion. *Nanotoxicology* **2008**, *2*, 144–154. [CrossRef]
57. Bianchi, M.G.; Campagnolo, L.; Allegri, M.; Ortelli, S.; Blosi, M.; Chiu, M.; Taurino, G.; Lacconi, V.; Pietroiusti, A.; Costa, A.L.; et al. Length-dependent toxicity of TiO_2 nanofibers: Mitigation via shortening. *Nanotoxicology* **2020**, *14*, 433–452. [CrossRef]
58. Porter, D.W.; Wu, N.; Hubbs, A.F.; Mercer, R.R.; Funk, K.; Meng, F.; Li, J.; Wolfarth, M.G.; Battelli, L.; Friend, S.; et al. Differential mouse pulmonary dose and time course responses to titanium dioxide nanospheres and nanobelts. *Toxicol. Sci.* **2013**, *131*, 179–193. [CrossRef]

Article

Toxicity to RAW264.7 Macrophages of Silica Nanoparticles and the E551 Food Additive, in Combination with Genotoxic Agents

Fanny Dussert [1], Pierre-Adrien Arthaud [1], Marie-Edith Arnal [1], Bastien Dalzon [2], Anaëlle Torres [2], Thierry Douki [1], Nathalie Herlin [3], Thierry Rabilloud [2] and Marie Carriere [1,*]

1 Université Grenoble-Alpes, CEA, CNRS, IRIG-DIESE, SyMMES, Chemistry Interface Biology for the Environment, Health and Toxicology (CIBEST), F-38000 Grenoble, France; fanny.dussert@cea.fr (F.D.); piarthaud@laposte.net (P.-A.A.); marie-edith.arnal@wanadoo.fr (M.-E.A.); thierry.douki@cea.fr (T.D.)

2 Chemistry and Biology of Metals, Université Grenoble Alpes, CNRS UMR5249, CEA, IRIG-DIESE-LCBM-ProMD, F-38054 Grenoble, France; bastien.dalzon@cea.fr (B.D.); Anaelle.torres@cea.fr (A.T.); thierry.rabilloud@cnrs.fr (T.R.)

3 Université Paris Saclay, CEA Saclay, IRAMIS NIMBE UMR 3685, 91191 Gif/Yvette CEDEX, France; nathalie.herlin@cea.fr

* Correspondence: marie.carriere@cea.fr; Tel.: +33-4-3878-0328

Received: 29 June 2020; Accepted: 16 July 2020; Published: 21 July 2020

Abstract: Synthetic amorphous silica (SAS) is used in a plethora of applications and included in many daily products to which humans are exposed via inhalation, ingestion, or skin contact. This poses the question of their potential toxicity, particularly towards macrophages, which show specific sensitivity to this material. SAS represents an ideal candidate for the adsorption of environmental contaminants due to its large surface area and could consequently modulate their toxicity. In this study, we assessed the toxicity towards macrophages and intestinal epithelial cells of three SAS particles, either isolated SiO_2 nanoparticles (LS30) or SiO_2 particles composed of agglomerated-aggregates of fused primary particles, either food-grade (E551) or non-food-grade (Fumed silica). These particles were applied to cells either alone or in combination with genotoxic co-contaminants, i.e., benzo[a]pyrene (B[a]P) and methane methylsulfonate (MMS). We show that macrophages are much more sensitive to these toxic agents than a non-differenciated co-culture of Caco-2 and HT29-MTX cells, used here as a model of intestinal epithelium. Co-exposure to SiO_2 and MMS causes DNA damage in a synergistic way, which is not explained by the modulation of DNA repair protein mRNA expression. Together, this suggests that SiO_2 particles could adsorb genotoxic agents on their surface and, consequently, increase their DNA damaging potential.

Keywords: silica; SiO_2; nanoparticle; E551; toxicity; genotoxicity; macrophage; intestine; co-exposure

1. Introduction

Synthetic amorphous silica (SAS) is an authorized food additive, known as E551 in the European Union. It is used for its anti-caking property in powdered food, including creamers, lyophilized soups, salt, and sugar [1]. It consists in particles with a primary diameter in the nano-range, i.e., lower than 100 nm, which aggregate to form large clusters with diverse morphologies [2]. This wide use has raised the concern of its safety and potential toxicity, in particular for the gastro-intestinal system. In the lung, inhalation exposure to SiO_2 has been reported to induce inflammation [3,4]. Moreover, risk assessment conducted with this substance and focused on the liver estimates a potential liver accumulation at the same level in humans and rodents in which adverse effects were found, suggesting that it could also be detrimental to human liver [5]. A recent review summarizing the literature relative to its safety assessment concludes in the absence of any relevant toxicity both at the systemic and local level after oral exposure [6]. The re-evaluation of this food additive by the EFSA Panel on Food Additives and Nutrient Sources added to Food (ANS), published in 2017, also concludes in the absence of toxic effects of E551 at the currently used levels, although silica was found to be absorbed through the gastro-intestinal tract and to accumulate in internal organs and the immune system. Synthetic amorphous silica are "generally recognized as safe" (GRAS) according to the US EPA. In particular, they were shown to induce only minor damage to DNA, which was considered to be within the normal physiological range [6]. Two recent reviews report that the literature relative to SiO_2-NP genotoxicity show inconsistent results, with some studies showing significant genotoxicity while others report the opposite [7,8].

Despite this apparent biocompatibility, combined effect of silica with other pollutants have been reported. The group of Zhiwei Sun described the impact of co-exposure of lung epithelial cells and zebrafish embryos to SiO_2-NPs with methylmercury or lead, as well as co-exposure to SiO_2-NPs and benzo[a]pyrene (B[a]P) on BEAS-2B bronchial epithelial cells, HUVEC endothelial cells, and zebrafish embryos [9–12]. These studies highlight increased cytotoxicity, apoptosis, oxidative stress, and inflammation in co-exposed cells, with both additive or synergistic effects of SiO_2 and the co-pollutant. The cardiovascular system is shown to be the main target organ where effects of these co-pollutants are observed. Synergistic interaction has also been reported between SiO_2-NPs and lead acetate in A549 alveolar epithelial cells, causing mitochondria-dependent apoptosis [13]. Moreover, cytotoxicity, oxidative stress, and apoptosis were reported for arsenic when co-exposed with SiO_2-NPs in HepG2 liver cells and fibroblasts [14]. Recently, Cao et al. reported increased cytotoxicity, oxidative stress, and translocation of the fungicide boscalid upon co-exposure of in vitro intestinal epithelial models to the E551 food additive (SiO_2), previously submitted to an in vitro simulated digestion [15]. All of these studies aiming at elucidating the impact of co-exposure of SiO_2 and environmental pollutants have been conducted on epithelial cells, endothelial cells, or fibroblasts. They show both additive or synergistic effect of SiO_2 and the co-contaminant, suggesting either SiO_2 particles acting as a cargo for the co-contaminant and facilitating its accumulation in cells, or a possible sensitization of cells towards the co-contaminant by SiO_2 particles.

Macrophages are major targets of SiO_2 in the organism, because they play a significant role in immunity and show particular sensitivity towards SiO_2-NPs [16,17]. However, systematic studies on the impact of co-exposure to SiO_2 and other pollutants, as well as studies on the genotoxicity of SiO_2 on this cell type are lacking, although we recently hypothesized that SiO_2-NPs could sensitize RAW264.7 macrophages towards DNA alkylating agents [17]. In this context, the aim of the present study was to compare the toxicity of SiO_2-NPs and the food additive E551 towards RAW264.7 macrophages and epithelial intestinal cells, with special focus on their genotoxicity and the impact of co-exposure with genotoxic pollutants. We chose two well-known genotoxic agents that cause DNA damage via different mechanisms, i.e., benzo[a]pyrene (B[a]P) and methane methylsulfonate (MMS). The rationale for these choices was that SiO_2-NPs are present in indoor air of some workplaces [18,19] while polycyclic aromatic hydrocarbon (PAHs) and among them B[a]P is an ubiquitous environmental pollutant, present in the atmospheric particulate matter as a consequence of incomplete combustion of organic matter as well as coal or petroleum distillation [20]. PAH are also present in the urban polluted atmosphere, sometimes in combination with inorganic NPs, like SiO_2, leading to co-exposure of the populations by inhalation. Last, PAHs are produced during cooking and SiO_2 is largely used as food additive [1]; therefore, co-exposure of the populations would also occur via ingestion. Exposure to B[a]P results in DNA strand breaks and adducts formed by benzo[a]pyrene-7,8-dihydrodiol-9,10-epoxide (BPDE), its most reactive metabolite. These two type of DNA damage, which can be detected with the comet assay and by HPLC-M/MS, respectively, are produced in the 1:10 ratio [21]. MMS is a typical model of N-alkylating agent that produces methylated bases in DNA, which are alkali-labile lesions that can be detected via the comet assay [22]. This model genotoxic agent was chosen because of our previously-published observation that SiO_2-NPs sensitized macrophages towards alkylating agents [17], with the aim of addressing the hypothesis that the E551 food additive would cause the same effect.

2. Materials and Methods

2.1. Chemicals and Reagents

Unless otherwise indicated, chemicals and reagents were >98% pure and were from Sigma–Aldrich. Silica particles were obtained from Sigma–Aldrich (Saint-Quentin Fallavier, France) (Ludox LS30, produced by Grace, and Fumed silica) or from an industrial collaborator producing food-grade precipitated silica (E551). Ludox LS30 was provided as a suspension, it was diluted in ultrapure water to reach the concentration of 1 mg/mL. Fumed silica and E551 were provided as powders, they were suspended in ultrapure water at the concentration of 1 mg/mL. They were not sonicated, because it would potentially degrade the structure of the food additive, which is primary particles aggregated as chaplets and then further agglomerated. These three particles were sterilized by heating at 80 °C overnight.

2.2. Cell Culture and Exposure

RAW 264.7 mouse macrophages and Caco-2 colorectal adenocarcinoma cells were obtained from the European Cell Culture Collection (ECACC, Salisbury, UK). RAW 264.7 were maintained at 37 °C, in a 5% CO_2 and 100% humidity incubator and grown in suspension, in non-adherent flasks, in RPMI 1640-Glutamax to which was added 50 U/mL of penicillin, 50 µg/mL streptomycin and 10% (*v*/*v*) fetal bovine serum (FBS). The cells were sub-cultured three times per week and then seeded at 200,000 cells per mL of growth medium. Caco-2 were maintained in DMEM Glutamax to which was added 1% nonessential amino-acids, 50 U/mL of penicillin, 50 µg/mL streptomycin, and 10% (*v*/*v*) FBS. HT29-MTX were kindly provided by Dr. T. Lesuffleur (INSERM) [23] and grown in the same medium as Caco-2 cells. Caco-2 and HT29-MTX cells were co-cultured at 75% Caco-2 and 25% HT29-MTX, as previously [24]. For acute exposure to particles, the cells were seeded in adherent plates, either 96-well (WST-1, trypan blue and LDH assay), 12-well (comet assay), or six-well (8-oxo-7,8-dihydro-2′-deoxyguanosine, 8-oxo-dGuo, measurement). In the acute exposure scheme, the cells were seeded at 500,000 cells per mL the day before exposure. They were exposed for 24 h to 10, 20, 50, or 100 µg/mL SiO_2 (WST1 and LDH assays), 10, 20, or 50 µg/mL SiO_2 (trypan blue cytotoxicity assay), or 10 µg/mL SiO_2 (comet assay and 8-oxo-dGuo measurement). In the repeated exposure scheme, the cells were seeded at 500,000 cells per mL and, then exposed 24 h later to 1 or 2 µg/mL SiO_2. Every second day during three weeks, the exposure medium was replaced with fresh medium containing 1 or 2 µg/mL SiO_2. This corresponds to nine successive exposures to SiO_2. At the end of this repeated exposure period, the cells were harvested with trypsin and seeded in clean plates, either 96-well (WST-1 assay), 12-well (comet assay), or six-well (Reverse transcription-quantitative polymerase chain reaction (RT-qPCR)). They were exposed 24 h later to 2, 5, 10, 25, or 50 µg/mL of fumed silica (WST1) or 10 µg/mL of fumed silica (comet assay, RT-qPCR).

2.3. Cytotoxicity Assays

Cytotoxicity was evaluated via the WST-1 assay (Roche, Mannheim, Germany), measuring cell metabolic activity and via staining with trypan blue (Sigma–Aldrich, Saint-Quentin Fallavier, France) and counting both viable cells (non-colored) and cells having impaired plasma membrane integrity (blue-colored cells). In the WST-1 assay, after the exposure period, the exposure medium was discarded and replaced by 100 µL of WST-1 diluted to the tenth, as indicated by the supplier. After 1.5 h of incubation at 37 °C, the quantification of metabolic activity was calculated from absorbance measurement at 450 nm, to which was subtracted background absorbance measured at 650 nm. The interference of SiO_2 particles with this assay was checked by centrifuging the plates, sampling

50 µL of each well and transferring it to a clean plate. Absorbance was then measured at 540 and 650 nm, and the obtained values were compared with those that were obtained before the centrifugation. The values were similar, we therefore considered that SiO_2 particles did not interfere with the WST1 assay. In the trypan blue assay, after the exposure period, the cells were harvested with trypsin-EDTA and trypan blue was applied to the cell suspension. Non-colored and blue cells were counted while using an automated cell counter (Countess, ThermoFisher Scientific, Illkirch, France). The absence of interference with the trypan blue assay was visually checked, by manually counting some samples and comparing the data with those that were obtained with the automatic counter. No significant difference was observed in cells that were exposed to 10–50 µg/mL SiO_2; however, at higher concentrations significant difference was observed, which were probably due to impaired detection of blue color or no color in cells having accumulated large quantities of NPs. For this reason, the results presented here were obtained at exposure concentrations that did not exceed 50 µg/mL. Cell membrane integrity was assessed using the Lactate dehydrogenase assay (Sigma–Aldrich, Saint-Quentin Fallavier, France), following the manufacturer's instructions, i.e., one volume of supernatant of exposed cells was sampled after the incubation period and mixed with two volumes of assay mix composed of equal proportions of assay substrate, cofactor, and dye. After incubation for 30 min at room temperature and in the dark, the reaction was stopped by adding 1/10 volume of 1N HCl and absorbance at 490 nm was measured. Triton X-100 (1%) was used as positive control. The absence of interference of SiO_2 particles with the assay was checked by centrifuging the supernatant of exposed cells, then measuring the absorbance at 490 nm and comparing it to the values obtained in samples that had not been centrifuged. We did not detect any significant difference, therefore we considered that SiO_2 particles did not interfere with the assay.

2.4. Genotoxicity Assays

DNA strand breaks and alkali-labile sites were assessed via the alkaline version of the Comet assay. At the end of the exposure period, cells were collected and stored at ~80 °C in sucrose (85.5 g/L), DMSO (50 mL/L) prepared in citrate buffer (11.8 g/L), pH 7.6. Ten thousand cells were mixed with 0.6% low melting point agarose (LMPA) and deposited on a slide that was previously coated with 1% agarose (n = 3). The cell/LMPA mix was allowed to solidify on ice for 10 min, then immersed in cold lysis buffer (2.5 M NaCl, 100 mM EDTA, 10 mM Tris, 10% DMSO, 1% Triton X-100, pH10) and incubated for 1 h at room temperature. The slides were then rinsed three times for 5 min in 0.4 M Tris pH 7.4. Subsequently, DNA was allowed to unwind for 30 min in the electrophoresis buffer (300 mM NaOH, 1 mM EDTA, pH > 13) and an electric field of 0.7 V/cm and 300 mA for 30 min was applied. Slides were neutralized in 0.4 M Tris pH 7.4 and stained with 50 µL of GelRed (Thermo Fisher Scientific, Illkirch, France). As positive control for the alkaline comet assay, 250 µM H_2O_2 was deposited onto the agarose layer containing the cells, and then incubated for 5 min on ice. Fifty comets per slide were analyzed while using Comet IV software (Perceptive Instruments, Suffolk, UK). The potential interference of SiO_2 nanoparticles with the comet assay have been assessed previously (for instance, see [25,26]). No significant interference was detected by Magdolenova et al. [26], while a slight overestimation of DNA damage is reported by Ferraro et al. In HeLa cells that were exposed for 48 h to 500 µg/mL SiO_2-NPs, but not to 50 or 200 µg/mL SiO_2-NPs [25]. In the present study, the cells were exposed to much lower concentrations of SiO_2 particles, and for shorter periods of time, we therefore considered that interference of SiO_2 particles with the comet assay is unlikely to be significant in our experimental conditions.

For quantification of modified DNA bases (HPLC-MS/MS), DNA was extracted as follows: the samples were extracted using DNeasy Blood and Tissue Kit (Qiagen, Les Ullis, France). They were homogenized in AL buffer from the kit, then proteinase K was added and the samples were incubated for 10 min at 56 °C. They were treated with RNase A for 2 min at room temperature and then loaded onto DNeasy Mini spin columns. After centrifugation, the samples loaded onto the columns were washed with AW1 buffer then with AW2 buffer. In the last step, DNA was eluted in 0.1 M deferoxamine to avoid spurious DNA oxidation. At this stage, the SiO_2 particles that were accumulated in cell cytoplasm are eliminated, because they are not eluted from the column. The samples were then digested for 2 h at 37 °C and pH 5.5 with a cocktail of enzymes (all purchased from Sigma–Aldrich, Saint-Quentin Fallavier, France) composed of phosphodiesterase II, DNase II, nuclease P1, and then for another 2 h at 37 °C, pH 8, with alkaline phosphatase and phosphodiesterase I. These samples were neutralized with HCl 0.1 N, filtered on 0.22 µm filter units to eliminate any remaining SiO_2 particles and injected onto the high performance liquid chromatography-tandem mass spectrometry system (HPLC/MS-MS). An API 3000 mass spectrometer (SCIEX, Villebon-sur-Yvette, France) was used in the multiple reaction monitoring mode with positive electrospray ionization. We monitored the m/z 284 $[M + H]^+ \rightarrow m/z$ 168 $[M + h\ \text{-}116]^+$ transition for the quantification of for 8-oxodGuo [27] and m/z 570 → 257 and m/z 570 → 454 for BPDE-N^2-dGuo [21]. A C18 reversed phase Uptisphere ODB column (Interchim, Montluçon, France) was used for chromatographic separations in an Agilent HPLC system (Agilent, Massy, France). The flow rate was 0.2 mL/min. The HPLC eluent was also analyzed using a UV detector set at 270 nm for the quantification of unmodified nucleosides. For the detection of 8-oxodGuo, elution was performed with a gradient of methanol in 2 mM ammonium formate, leading to a retention time of around 29 min. For the quantification of BPDE-N^2-dGuo, the HPLC mobile phase was a gradient of 6 to 80% of acetonitrile in 2 mM ammonium formate (pH 6). The retention time of the BPDE-N^2-dGuo adduct was 24.5 min. For both 8-oxodGuo and BPDE-N^2-dGuo measurements, results were expressed in the number of adducts per million normal bases. Because no SiO_2 particle was injected in the columns, we consider that SiO_2 particles could not interfere with the measurements.

2.5. RT-qPCR

RNA from exposed cells was extracted using the GenElute™ mammalian total RNA miniprep kit (Sigma–Aldrich, Saint-Quentin Fallavier, France) with the optional DNAse treatment step and reverse-transcribed to complementary DNA (cDNA) while using the SuperScript III Reverse Transcriptase kit (Thermo Fisher Scientific, Illkirch, France), according to the manufacturers' protocols. The first step in this assay consists in the elimination of any cell debris and material via filtration on a column. We consider that SiO_2 particles must be retained on this column and, therefore, are unlikely to interfere with the following stages of mRNA extraction, RT, and qPCR. RNA concentration and purity were assessed by measuring A260/A280 and A260/A230 absorbance ratios using a Nanodrop ND-1000 (Thermo Fisher Scientific, Illkirch, France). For the qPCR, cDNA from each of the three biological replicates of each exposure condition was loaded in duplicate on a 96-well qPCR plate. qPCR was performed on a CFX96 thermocycler (Biorad, Marne-la-coquette, France) using the following thermal cycling steps: 95 °C for 5 min, then 95 °C for 15 s, 55 °C for 20 s and 72 °C for 40 s 40 times, and finally 95 °C for 1 min, 55 °C for 30 s and 95 °C for 30 s for the dissociation curve. Cq was determined by the CFX96 Manager (Biorad, Marne-la-coquette, France) used with default settings. Glyceraldehyde-3-phosphate dehydrogenase (GAPDH) and 18S ribosomal 1 (S18) were chosen as reference genes for normalization, and validated while using the BestKeeper tool, version 1 [28]. mRNA expression analysis, normalization, and statistical analysis were performed with REST 2009 software [29], which uses the $\Delta\Delta$Cq method and a pair-wise fixed reallocation randomization test. The PCR efficiencies were experimentally checked for compliance using a mix of all samples, with a quality criterion of 2 ± 0.3.

2.6. Statistical Analysis

Each experiment was repeated at least three times independently. The statistical tests were performed using the Statistica software (version 7.1, Statsoft, Chicago, IL, USA). As normality assumptions for valid parametric analyses were not satisfied, a non-parametric test was used, i.e., Kruskal–Wallis one-way analysis of variance. When significance was demonstrated, paired comparisons were performed using Mann–Whitney tests. The results were considered to be statistically significant (*) when the p value was <0.05.

3. Results

3.1. Physico-Chemical Characterization of SiO_2 Particles

The three SiO_2 particles were prepared as suspensions in water and sterilized by pasteurization. Their size distribution analysis showed the agglomeration of E551, which formed agglomerates of particles with diameter >2 µm. Conversely, Fumed silica and LS30 formed stable suspensions with mean hydrodynamic diameters of 270 and 24 nm, respectively (Figure 1a). Polydispersity indexes were 0.28, 0.21, and 0.24 for E551, Fumed SiO_2, and LS30, respectively. Their zeta potential was slightly negative, with values between ~10 and ~30 mV (Figure 1b), which suggested a tendency towards agglomeration. In RAW 264.7 exposure medium, all three particles agglomerated. Fumed silica and LS30 still formed stable suspensions, with hydrodynamic diameters of 1203 and 381 nm, respectively, while E551 formed very large agglomerates with diameter >5 µm. The values were similar in Caco-2/HT29-MTX exposure medium (not shown). As expected, the TEM images showed that E551 (Figure 1c) and Fumed SiO_2 (Figure 1d) were composed of aggregates of fused nanoparticles with primary diameter of 15–20 nm, while LS30 was composed of SiO_2 nanoparticles with average primary diameter 14.3 ± 2.2 nm, as measured from 100 particles on TEM images (Figure 1e).

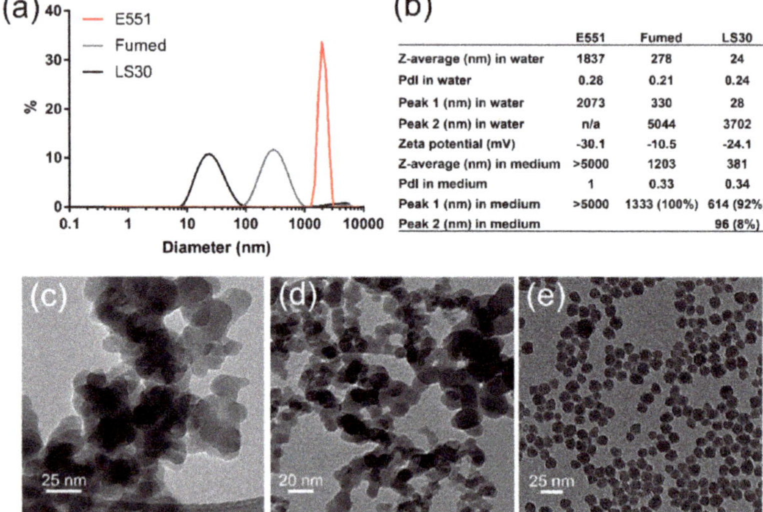

	E551	Fumed	LS30
Z-average (nm) in water	1837	278	24
PdI in water	0.28	0.21	0.24
Peak 1 (nm) in water	2073	330	28
Peak 2 (nm) in water	n/a	5044	3702
Zeta potential (mV)	-30.1	-10.5	-24.1
Z-average (nm) in medium	>5000	1203	381
PdI in medium	1	0.33	0.34
Peak 1 (nm) in medium	>5000	1333 (100%)	614 (92%)
Peak 2 (nm) in medium			96 (8%)

Figure 1. Physico-chemical characterization of SiO_2 particles. (**a**) Size distribution of SiO_2 particles in water; (**b**) hydrodynamic diameter, polydispersity index and zeta potential for SiO_2 particles dispersed in water; (**c**) TEM image of E551; (**d**) TEM image of Fumed silica; and, (**e**) TEM image of LS30.

3.2. Acute Cytotoxicity and Genotoxicity of SiO_2 Particles

First, cell viability was assessed on RAW264.7 macrophages and Caco-2/HT29-MTX exposed to the three SiO_2 particles. All three SiO_2 particles altered RAW 264.7 cell viability after acute exposure for 24 h (Figure 2). In the WST1 assay, Fumed silica altered more intensely cell metabolic activity than E551, which itself altered more intensely cell metabolic activity than LS30 (Figure 2a). Using the trypan blue assay, the three particles showed similar cytotoxicity, which was lower than cell metabolic activity alteration (Figure 2b). Conversely, in Caco-2/HT29-MTX cells, no significant reduction of cell metabolic activity (WST1 assay) and cell membrane integrity (LDH assay) were detected (Figure 2c,d), respectively.

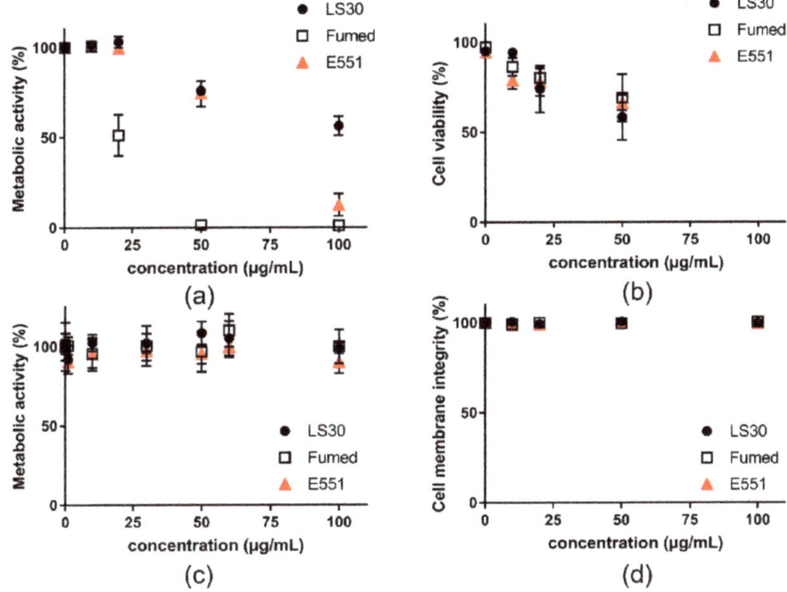

Figure 2. Cytotoxicity of SiO_2 particles, acute exposure for 24 h to SiO_2 particles. (**a**) metabolic activity of RAW264.7 cells assessed via the WST-1 assay; (**b**) cytotoxicity assessed in RAW264.7 cells via trypan blue staining; (**c**) metabolic activity of Caco-2/HT29-MTX cells assessed via the WST-1 assay; and, (**d**) membrane integrity of Caco-2/HT29-MTX cells assessed via the LDH assay. Mean ± standard deviation of five replicates (n = 5).

This experiment allowed the determination of the dose to be applied in genotoxicity assays, particularly in the comet assay, which should be a concentration leading to less than 20–30% of cell death. With respect to these results, the genotoxicity assays were performed on RAW264.7 cells that were exposed to 10 µg/mL SiO_2, because it was the highest tested concentration causing no significant cell death in both WST-1 and trypan blue assays. For Caco-2HT29-MTX, since no cytotoxicity of SiO_2 particles was observed, even at the highest concentrations tested, we chose to expose cells to 5, 15, and 30 µg/mL SiO_2 particles, as suggested to avoid any interference with the assays [30].

At these sub-lethal concentrations, the three SiO_2 particles significantly increased the number of strand breaks and/or alkali-labile sites in RAW264.7 cells, in the alkaline comet assay (Figure 3a), which probes their capacity to induce oxidative damage to DNA. The level of DNA damage was similar in cells that were exposed to Fumed silica and E551; it was significantly higher than the level of DNA damage that was caused by LS30. Conversely, none of these SiO_2 particles increased the level of 8-oxo-dGuo in the DNA of exposed cells (Figure 3b). In Caco-2/HT29-MTX cells, none of the particles induced any increase of DNA damage in exposed cells, neither in the comet assay nor via direct measurement of 8-oxo-dGuo by HPLC-MS/MS (Figure 3c,d), respectively.

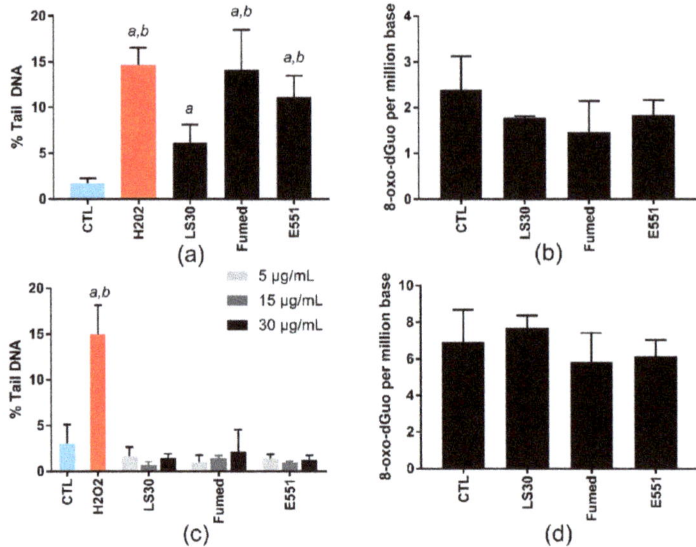

Figure 3. Genotoxicity of SiO_2 particles, acute exposure. (**a**,**c**) DNA strand breaks and/or alkali-labile sites, assessed via the alkaline comet assay, in RAW264.7 cells exposed to 10 µg/mL SiO_2 (**a**) and in Caco-2/HT29-MTX cells exposed to 5, 15, or 30 µg/mL SiO_2 (**c**). H_2O_2 (250 µM) was used as positive control. Mean ± standard deviation of five independent experiments (n = 5); (**b**,**d**) quantification of 8-oxo-dGuo by high performance liquid chromatography-tandem mass spectrometry system (HPLC-MS/MS), in RAW264.7 cells exposed to 10 µg/mL SiO_2 (**b**) and Caco-2/HT29-MTX cells exposed to 50 µg/mL SiO_2 (**d**). Mean ± standard deviation of three independent replicates from the same experiment (n = 3). Statistical significance, ^{a} $p < 0.05$, exposed vs. CTL (untreated cells), ^{b} $p < 0.05$, exposed vs. LS30.

3.3. Cytotoxicity and Genotoxicity of LS30, Fumed Silica or E551 after Repeated Exposure

Because SiO_2 particles only showed toxic outcomes in RAW264.7, we then focused on this cell line in the following experiments. The cells were repeatedly exposed to 1 or 2 µg/mL of these SiO_2 particles in order to assess the hypothesis of progressive accumulation of SiO_2 particles in RAW264.7 leading to additive level of DNA damage. This protocol mimics long term exposure to SiO_2 particles via ingestion. Cells were seeded in adherent plates and exposed nine times to these concentrations of SiO_2 particles at the frequency of one exposure every two days. This corresponds to three weeks of exposure and the cumulative dose was 9 and 18 µg/mL, respectively. This repeated exposure did not cause any overt alteration of cell metabolic activity (Figure 4a). For comparison, acute exposure to 10 µg/mL or 20 µg/mL SiO_2 particles also did not decrease cell viability, except in RAW 264.7 cells that were exposed to 20 µg/mL Fumed SiO_2, which led to ~50% decrease of cell metabolism (see Section 3.2). Conversely, this repeated exposure induced DNA strand breaks and/or alkali-labile sites, which increased with exposure concentration (Figure 4b). At each exposure concentration, all three SiO_2 particles produced the same level of DNA damage in RAW 264.7 cells. The level of DNA damage in cells repeatedly exposed to 1 µg/mL SiO_2 particles (cumulative concentration: 9 µg/mL SiO_2) was slightly less intense than in cells that were acutely exposed to 10 µg/mL SiO_2 (see Section 3.2). Conversely, in cells repeatedly exposed to 2 µg/mL SiO_2 particles (cumulative concentration: 18 µg/mL SiO_2), the level of DNA damage was much higher, i.e., close to 50% Tail DNA; with high variability.

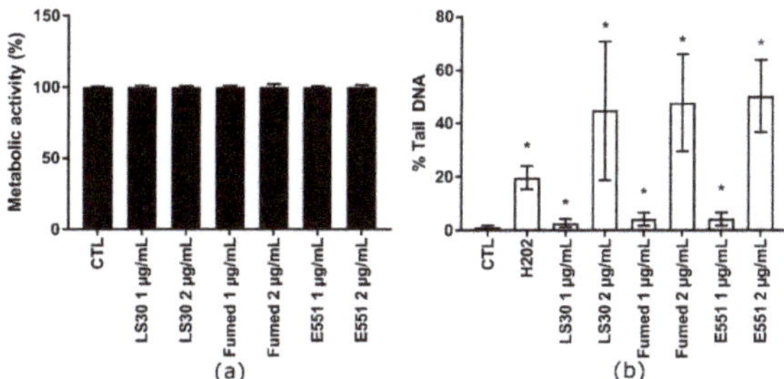

Figure 4. Cytotoxicity of SiO_2 particles, repeated exposure. RAW 264.7 were exposed repeatedly to 1 or 2 µg/mL SiO_2 particles, three times per week for 3 weeks. (**a**) Metabolic activity of RAW264.7 cells measured using the WST-1 assay and (**b**) genotoxicity of SiO_2 particles assessed via alkaline comet assay. Mean ± standard deviation of five replicates (WST-1) and three comet experiments performed independently, with three slides per experiment. Statistical significance: * $p < 0.05$, CTL vs. exposed.

We then investigated the potential of each of these SiO_2-NP to modify the effects of the most genotoxic NP studied here, namely Fumed silica (see Section 3.2). This mimics a situation where SiO_2 is chronically ingested every day, and then one day of intense pollution a significant amount of SiO_2 is inhaled, then ingested acutely due to mucociliary clearance. Cells were repeatedly exposed to 1 µg/mL of SiO_2-NPs three times per week for 3 weeks and then subsequently acutely exposed to 10 µg/mL of Fumed silica. In the cytotoxicity assay, the preliminary repeated exposure to SiO_2 did not significantly change the overall response to subsequent acute exposure to Fumed silica (Figure 5a). In contrast, in the genotoxicity assay, the level of DNA damage caused by the acute exposure to Fumed silica was significantly increased in cells that had been previously exposed to 1 µg/mL SiO_2, as compared to cells not previously exposed to SiO_2 (Figure 5b, white bars). When adding the level of damage caused by the repeated exposure to SiO_2 particles to that caused by a single acute exposure to 10 µg/mL of Fumed SiO_2 (without pre-exposure), the obtained value was similar to that observed in cells that were repeatedly exposed to SiO_2 and then acutely to Fumed SiO_2 (Figure 5b, grey bars). This suggest progressive, cumulative accumulation of SiO_2 in repeatedly- and then acutely-exposed cells, resulting in additive levels of DNA damage.

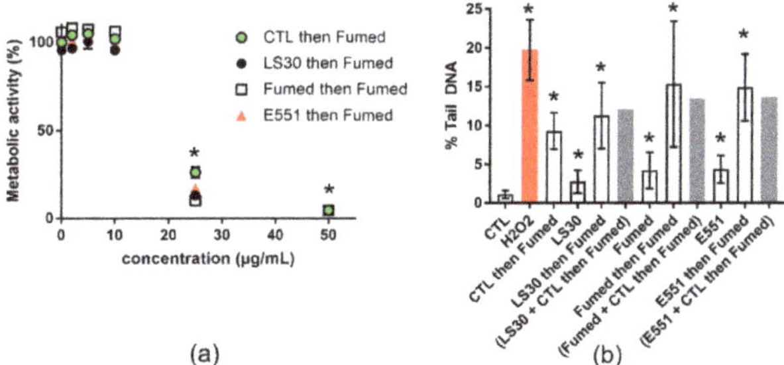

Figure 5. Cyto- and genotoxicity of SiO_2 particles, repeated exposure followed by acute exposure to Fumed SiO_2. (**a**) Cell metabolic activity impairment after repeated exposure to 1 µg/mL SiO_2 particles (CTL or LS30 or Fumed SiO_2), followed by 24 h exposure to 2, 5, 10, 25, or 50 µg/mL of Fumed silica. (**b**) Genotoxicity assessed via alkaline comet assay, on RAW 264.7 cells exposed repeatedly to 1 µg/mL SiO_2 particles, three times per week for three weeks, followed by a single acute exposure to 10 µg/mL Fumed SiO_2 for 24 h. Mean ± standard deviation of five replicates (WST-1) and three comet experiments performed independently, with three slides per experiment. Statistical significance: * $p < 0.05$, CTL vs. exposed.

3.4. Cytotoxicity and Genotoxicity after Co-Exposure to SiO_2 and B[a]P or MMS

We then tested the hypothesis of sensitization of RAW264.7 macrophages towards genotoxic agents by SiO_2-NPs. RAW264.7 were acutely co-exposed to SiO_2 particles and known genotoxic agents. First, they were exposed to 0.2–2 µM B[a]P or a mixture of 0.2–2 µM of B[a]P and 10 µg/mL SiO_2 particles. In these conditions, no significant modulation of cell viability was detected (Figure 6a) up to 2 µM of B[a]P, which is very high as compared to environmentally-relevant concentrations. Indeed, the values measured in the bloodstream of contaminated people are in the range of some nm, and they can reach 1 µM in some industrial sectors. The main damage to DNA caused by B[a]P are DNA-BPDE adducts and DNA strand breaks, the former being much more frequent than the latter [31]. Analysis of the DNA extracted from cells exposed to either 0.2–2 µM B[a]P or a mixture of 0.2–2 µM of B[a]P and 10 µg/mL SiO_2 particles revealed the absence of BPDE adducts. Figure 6b show the retention time of BPDE-N^2-dGuo in HPLC-MS/MS (Figure 6b, blue chromatogram), the spectrum obtained from RAW264.7 cells exposed to 2 µM B[a]P (Figure 6b, red chromatogram) or to 2 µM B[a]P and 10 µg/mL Fumed SiO_2 for 24 h (Figure 6b, green chromatogram), showing no evidence of a BPDE-N^2-dGuo peak. The spectra obtained from cells co-exposed to 0.2–2 µM of B[a]P and 10 µg/mL LS30, Fumed SiO_2 or E551 were similar.

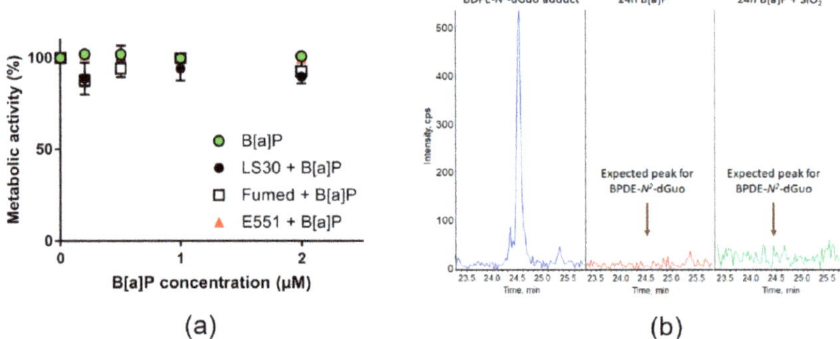

Figure 6. Cyto- and genotoxicity of B[a]P or SiO_2 co-exposed with B[a]P towards RAW264.7 cells. (**a**) Cell metabolic activity, assessed via the WST1 assay; (**b**) HPLC-MS/MS chromatograms obtained upon quantification of BPDE adduct to DNA showing the expected position of the BPDE adduct peak (BPDE-N^2-dGuo) (blue, left), of DNA extracted from cells exposed to 2 µM B[a]P (red, middle) and of DNA extracted from cells exposed to 2 µM B[a]P and 10 µg/mL Fumed SiO_2 (green, right).

The cells were directly exposed to BPDE then their DNA was extracted and the presence of BPDE-N^2-dGuo adducts was monitored by HPLC-MS/MS in order to verify that this absence of BPDE adducts reported in Figure 6 did not result from the incapacity of RAW264.7 cells to metabolize B[a]P to BPDE. Again, no BPDE-N^2-dGuo adduct was detected (not shown), confirming that the absence of BPDE adducts in cells B[a]P-exposed was not due to a lack of metabolization of B[a]P.

The cells were then exposed to MMS or a mixture of MMS and 10 µg/mL SiO_2. The cytotoxicity of all three SiO_2 particles, when cells were co-exposed to 10 µg/mL SiO_2 particle and 100–500 µM MMS, did not differ from cytotoxicity of the corresponding concentration of MMS (Figure 7a). As expected, MMS induced a dose-dependent elevation of DNA strand breaks and/or alkali-labile sites in RAW264.7 cells in the alkaline comet assay (Figure 7b). When considering cells that were exposed to a mixture of 10 µg/mL SiO_2 particle and 100 µM MMS, the level of DNA damage was greater, as compared to cells exposed to 100 µM MMS (Figure 7b). We then added the level of damage observed in cells exposed to 10 µg/mL SiO_2 to that observed in cells that were exposed to 100 µM MMS, while assuming that the effect of these two toxic agents could be purely additive. When comparing these calculated values with the experimental values that were obtained upon co-exposure to SiO_2

and MMS, the experimental values were greater than the calculated values (Figure 7c), suggesting a synergistic interaction between SiO_2 particles and MMS.

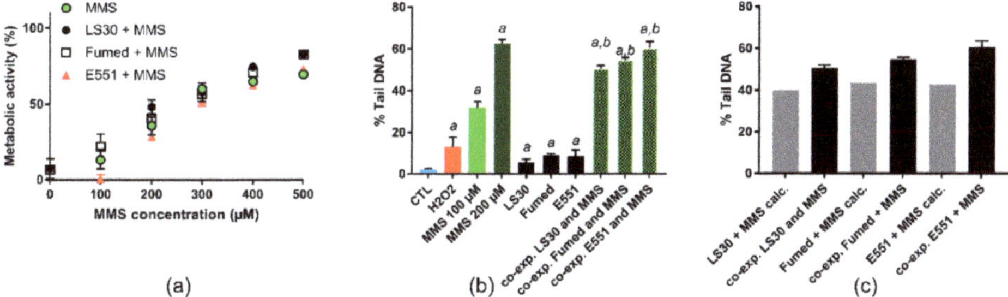

Figure 7. Cyto- and genotoxicity on RAW264.7 cells of MMS or co-exposure to MMS and SiO_2. (**a**) Cell metabolic activity, assessed via the WST1 assay; (**b**) DNA strand breaks and/or alkali-labile sites assessed via the alkaline comet assay in RAW264.7 cells exposed to 10 µg/mL SiO_2. Positive control: H_2O_2 (250 µM); (**c**) comparison of experimental results (level of DNA damage in cells co-exposed to MMS and SiO_2) and calculated values (level of DNA damage in cells exposed to MMS + level of DNA damage in cells exposed to SiO_2) WST1: mean ± standard deviation of five independent experiments (n = 5); comet assay: mean ± standard deviation of two independent experiments with three slides per experiment (n = 2). Statistical significance, *a*: $p < 0.05$, exposed vs. CTL (untreated cells), *b*, $p < 0.05$, co-exposed to SiO_2 particle and MMS vs. exposed to the respective SiO_2 particle.

A hypothesis for explaining this synergistic interaction between SiO_2 particles and MMS would be that SiO_2 particles would impair DNA repair activities in exposed cells. To test this hypothesis, the mRNA expression of genes encoding DNA repair proteins were analyzed in cells that were exposed to LS30, Fumed SiO_2, and E551. MMS is an alkylating agent, which mainly methylates N7-deoxyguanosine and N3-deoxyadenosine. These methylated bases are unstable and are rapidly hydrolyzed into an abasic site. Damage that is caused by MMS is repaired via the base-excision repair (BER) pathway and DNA methyltransferases [32,33]. We measured the mRNA expression of DNA repair enzymes involved in the BER pathway, namely the endonuclease APE1, XRCC1, and PARP1 that coordinate the resynthesis and polymerization steps of BER (for more detail on this DNA repair pathway, see [33]). No significant modulation of mRNA expression of these three proteins was observed (Figure 8), which suggested that this DNA repair pathway was not affected by exposure to SiO_2 particles.

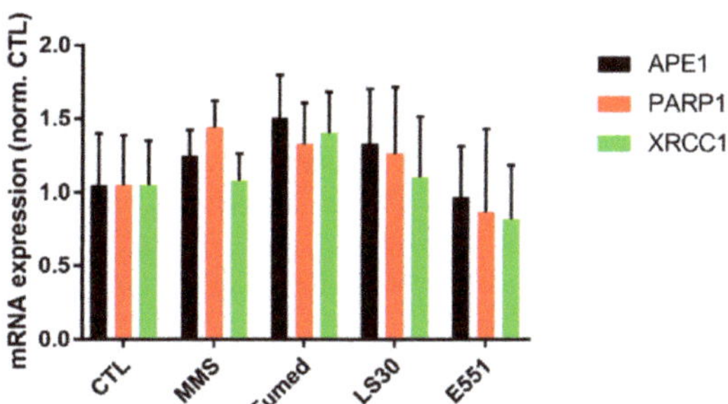

Figure 8. mRNA expression of three proteins involved in DNA repair via the base-excision pathway. Mean ± standard deviation of two independent experiments with three slides per experiment (n = 2). Statistical significance, none of the conditions induced statistically significant changes ($p > 0.05$, exposed vs. CTL).

4. Discussion

In this study, we compared the toxicity of synthetic amorphous silica in RAW264.7 macrophages and in Caco-2/HT29-MTX epithelial intestinal cells, and showed that RAW264.7 are more sensitive to SiO_2 than these epithelial cells, with greater impact on cell viability, as assessed via the WST1 assay and a higher level of DNA damage in the comet assay. This greater sensitivity of macrophages has already been described in several reports, e.g., in [17]. It is certainly related to their capacity to accumulate larger quantities of particles as compared to epithelial cells. This is explained by their physiological function, which is the scavenging of exogenous matter in the body, especially large-sized material, such as bacteria and viruses. The size of SiO_2 agglomerates to which RAW264.7 cells have been exposed in the present study, particularly Fumed SiO_2 and E551, falls within the optimal size range of material that is efficiently phagocytosed by macrophages, i.e., 2–3 μm [34]. This suggests that intracellular accumulation of SiO_2 in RAW 264.7 cells would be intense, while accumulation in intestinal epithelial cells, which are not phagocytosis-competent, would be less efficient, as it mainly derives from endocytosis [35]. Moreover, we used here a Caco-2/HT29-MTX co-culture, in which HT29-MTX cells produce some protective mucus [23,36] and this could also explain their resistance to SiO_2 particles, also owing to lower intracellular accumulation due to the entrapment of particles in mucus.

The three SiO_2 particles used here vary in their physico-chemical properties. LS30 is composed of isolated nanoparticles, while Fumed SiO_2 and E551 are composed of constituent nanoparticles fused together to form large chaplets of particles. Moreover, Fumed SiO_2 is a pyrogenic silica, while LS30 and the E551 used in this study are both produced by a wet process (i.e., precipitated SiO_2). We observed that Fumed SiO_2 shows greater toxicity than E551 and LS30, which is in line with the literature [37,38]. The greater toxicity of pyrogenic silica has been related to their higher surface reactivity [37], which could be explained by the presence of strained three-membered rings, to higher hydroxyl content and chainlike agglomeration [38]. Here, the cytotoxicity data obtained with the three SiO_2 confirm these hypotheses, with LS30 non-aggregated colloidal SiO_2 being the least toxic, followed by E551 aggregated SiO_2 (and synthesized as precipitated SAS), and then finally Fumed SiO_2, which is aggregated and pyrogenic. Regarding their genotoxicity, E551 and Fumed SiO_2 do not show a significant difference, but LS30 is less prone to damaging DNA. This could be the basis of future recommendation on the physico-chemistry of SiO_2 that are authorized as food additive, with possibly the suggestion of reducing, as much as possible, the structural defects in pyrogenic SiO_2 in order to reduce their toxicity.

Our initial objective was to evaluate whether co-exposure to SiO_2 and genotoxic agents could modulate the DNA damaging potential of genotoxic agents. Indeed, SiO_2 could adsorb metals or environmental pollutants on their surface and facilitate their accumulation in cells or organisms. This would increase their toxicity. In contrast, the adsorption of some pollutants on the surface of SiO_2 particles could inactivate these co-contaminants by modifying their configuration [39] or could reduce their availability, therefore reducing their toxicity. Moreover, we previously reported that RAW264.7 macrophages were more sensitive to the alkylating agent styrene oxide when previously exposed to SiO_2 nanoparticles [17]. We observe increased genotoxicity of MMS when co-exposed to RAW264.7 macrophages with the three SiO_2 particles, and the interaction between MMS and SiO_2 was synergistic rather than simply additive. We attempted similar experiments with B[a]P, but could not detect any genotoxic potential of this substance in RAW 264.7 cells. The synergistic interaction of SiO_2 particles with co-contaminants has been reported in various studies, particularly with metals, such as cadmium, methylmercury, arsenic, and lead [13,14,40–44], or with B[a]P [10–12]. Increased genotoxic potential has been highlighted for SiO_2 and B[a]P in epithelial cells [10,11,44]. The authors used the comet assay to assess DNA damage that is caused by B[a]P and, therefore, only detected strand breaks, which have been shown to be produced in lower amount as compared to the BPDE-DNA adducts in hepatocytes and lung cells [21,31]. Moreover, the sensitivity of BPDE-DNA adducts is largely higher than the sensitivity of the comet assay. Some BPDE-DNA adducts can be detected and quantified in cells exposed to some tens of femtomolars of B[a]P, while the exposure concentration should be higher than 1 μM to be able to detect some DNA damage in the comet assay [31]. A hypothesis to explain that we did not detect any DNA-BPDE adducts in RAW 264.7 macrophages would be that these cells do not have the capacity to metabolize B[a]P to BPDE. However, the RAW 264.7 cell line has been shown to express active P450 cytochromes, which are responsible for this metabolization [45]. Moreover, we did not observe any BPDE-DNA adduct, even when RAW 264.7 cells were directly exposed to BPDE. One possible reason to explain the absence of DNA-BPDE adducts would be that BPDE is quickly expulsed out of RAW264.7 cells before being able to reach the cell nucleus and to damage DNA. Another explanation could be a much more active phase II detoxification pathways leading a complete conversion of BPDE into excreted metabolites. The situation may be different in epithelial and endothelial cells, which have very different metabolisms as compared to macrophages. Macrophages, whose physiological function is to clean the organism from toxic substances and materials, certainly shows a greater ability to discard such metabolites. The genotoxic impact of B[a]P and SiO_2 was shown to be synergistic in HUVEC cells in the study by Otieno-Asweto et al. [10] and additive in BEAS-2B cells the study by Wu et al. [11]. The major differences between these two studies are i) the cell line on which the assays were conducted and ii) the applied concentrations. BEAS-2B had been exposed to 5 μg/mL SiO_2-NPs and 5 μM B[a]P, while HUVECs had been exposed to 10 μg/mL SiO_2 and 1 μM B[a]P. Both of the cell lines were exposed in the same medium. The use of different exposure concentrations does not allow for direct comparison of the results; still, HUVECs cells respond more intensely to these toxic agents than BEAS-2B cells, which suggests that they are more sensitive. In both studies, the rate of apoptosis in cells co-exposed to SiO_2 and B[a]P was significant, i.e., 50% of apoptosis rate in co-exposed cells compared to 25% in control HUVECs cells (approximately 20% of cell death in the CCK-8 assay) [10] and approximately 25% in co-exposed cells as compared to 15% in control BEAS-2B cells (<10% of cell death in the CCK-8 assay) [11]. Because the authors do not measure BPDE-DNA adducts, but rather use the comet assay to assess the genotoxicity of B[a]P, one can hypothesize that the DNA damage detected in these studies could be rather an indirect measurement of DNA fragmentation occurring when cells undergo apoptosis, than a direct impact of SiO_2 and B[a]P on DNA. It could also derive from the oxidative stress that results from HAP metabolization. The authors do not propose any hypothesis that could explain the either synergistic of additive interaction of SiO_2 and B[a]P in the genotoxicity experiments. One explanation could be that SiO_2 particles impair DNA repair processes. This hypothesis is not supported by our mRNA expression experiments. However, DNA repair processes function on the basis of already existing DNA repair proteins, so mRNA expression measurement is perhaps not a reliable method for assessing DNA repair activity in these cells. Importantly, we previously detected a decrease in the level of proteins related to the nucleotide excision repair pathway (NER) in RAW 264.7 cells that were exposed to a colloidal silica NPs with similar characteristics as LS30 [17]. We could hypothesize that a similar mechanism could be at play here, with the BER pathway. Unfortunately, such low amplitude changes, although putatively biologically significant, are technically difficult to detect. One could also hypothesize that SiO_2 particles adsorb large amounts of MMS on their surface and act as a vector to transport it inside cells, thereby increasing the overall level of cell exposure to this genotoxic agent. In this model, silica behaves as an adsorptive material, such as when it is used in chemistry as a chromatographic support. In this frame, the medium hydrophilicity of MMS (LogKow = ~0.87) makes it a good candidate for an adsorption–release mechanism on silica, where the much more hydrophobic B[a]P (LogKow = 6.1) will not adsorb appreciably on silica in a complex environment, where more hydrophobic macromolecules (e.g., proteins) are present.

When considering the experiment where cells were subjected to repeated exposure to SiO_2 particles followed by acute exposure to Fumed SiO_2, the level of DNA damage measured in cells is exactly the cumulative level calculated by adding the level of damage after repeated exposure to the level of damage after acute exposure. This suggests that the genotoxicity of SiO_2 particles towards RAW264.7 cells is cumulative, certainly deriving from progressive accumulation of SiO_2 particles that is not compensated by any exocytosis of the particles out of cells. This confirms the previously observed trend of progressive accumulation of fluorescent SiO_2-NPs in this cell line [46].

SiO_2 particles are generally considered to be non-toxic, especially non-genotoxic, although several recent studies show their potency to cause DNA damage, as assessed via the comet assay, micronucleus assay, and gene mutation assay (for instance, see [47–50]). Here, we show significant damage to DNA caused by the three SiO_2 particles on macrophages. The rationale for testing nanoparticle genotoxicity on macrophages can be questioned, because genotoxicity is classically assessed as a preliminary event leading to gene mutation, which may then lead to cancer. However, cancers that are linked to macrophages, i.e., histiocytomas, are very rare. However macrophages are interesting tools to study particle toxicity because they heavily accumulate particles. Therefore, they could serve as model cells to predict the hazard dimension of particle genotoxicity.

5. Conclusions

In this article we show that SiO_2 particles cause genotoxic damage in the RAW 264.7 cell line, but not in a co-culture of Caco-2 and HT29-MTX intestinal epithelial cells, which confirms the particular sensitivity of macrophages towards SiO_2 that has already been observed elsewhere. The genotoxic damage is significant whatever the SiO_2 physico-chemical properties and purity, i.e., either with isolated nanoparticles or with agglomerated-aggregates of fused primary particles and either with food-grade SiO_2 or with non-food-grade SiO_2. The level of DNA damage increases linearly upon repeated exposure, which suggests progressive accumulation of particles into cells causing progressive elevation of the level of DNA lesions, and no release of particles from cells. While B[a]P do not induce any DNA damage in RAW 264.7 cells, SiO_2 particles and MMS synergistically induce the elevation of DNA damage in exposed cells. This synergistic effect is not correlated with any significant modulation of DNA repair in exposed cells. Taken together, these data suggest that SiO_2 particles could serve as cargo for genotoxic agents, therefore increasing their DNA damaging potential.

Author Contributions: Data curation, M.C.; Formal analysis, M.C.; Funding acquisition, T.R. and M.C.; Investigation, F.D., P.-A.A., M.-E.A., B.D., A.T., T.D., N.H. and M.C.; Methodology, T.R. and M.C.; Project administration, T.R. and M.C.; Resources, T.R. and M.C.; Supervision, T.R. and M.C.; Writing—original draft, M.C.; Writing—review & editing, M.-E.A., T.D. and T.R. All authors have read and agreed to the published version of the manuscript.

Funding: This research was funded by Excellence Initiative of Aix-Marseille University—A*MIDEX, a French "Investissements d'Avenir" program (grant number ANR-11-IDEX-0001-02), through its associated Labex SERENADE project (grant number ANR-11-LABX-0064). This work was funded by the French National Research Program for Environmental and Occupational Health of ANSES (PNREST 2015/032, Silimmun Grant) and the French National Research Agency (ANR-16-CE34-0011, PAIPITO grant).

Acknowledgments: This work used the platforms of the Grenoble Instruct centre (ISBG; UMS 3518 CNRS-CEA-UJF-EMBL) with support from FRISBI (ANR-10-INSB-05-02) and GRAL (ANR-10-LABX-49-01) within the Grenoble Partnership for Structural Biology (PSB). The IBS electron microscope facility is supported by the Rhône-Alpes Region, the Fondation Recherche Medicale (FRM), the fonds FEDER, the Centre National de la Recherche Scientifique (CNRS), the CEA, the University of Grenoble, EMBL, and the GIS-Infrastrutures en Biologie Sante et Agronomie (IBISA). The authors particularly thank Daphna Fenel and Guy Schoehn for providing access to this infrastructure. We also thank Olivier Renard for providing access to DLS equipment at the CEA/LVME laboratory.

Conflicts of Interest: The authors declare no conflict of interest.

References

1. Dekkers, S.; Krystek, P.; Peters, R.J.; Lankveld, D.P.; Bokkers, B.G.; van Hoeven-Arentzen, P.H.; Bouwmeester, H.; Oomen, A.G. Presence and risks of nanosilica in food products. *Nanotoxicology* **2011**, *5*, 393–405. [CrossRef] [PubMed]
2. De Temmerman, P.J.; Van Doren, E.; Verleysen, E.; Van der Stede, Y.; Francisco, M.A.; Mast, J. Quantitative characterization of agglomerates and aggregates of pyrogenic and precipitated amorphous silica nanomaterials by transmission electron microscopy. *J. Nanobiotechnol.* **2012**, *10*, 24. [CrossRef] [PubMed]

3. Arts, J.H.; Muijser, H.; Duistermaat, E.; Junker, K.; Kuper, C.F. Five-day inhalation toxicity study of three types of synthetic amorphous silicas in Wistar rats and post-exposure evaluations for up to 3 months. *Food Chem. Toxicol.* **2007**, *45*, 1856–1867. [CrossRef] [PubMed]
4. Sun, B.; Wang, X.; Liao, Y.P.; Ji, Z.; Chang, C.H.; Pokhrel, S.; Ku, J.; Liu, X.; Wang, M.; Dunphy, D.R.; et al. Repetitive Dosing of Fumed Silica Leads to Profibrogenic Effects through Unique Structure-Activity Relationships and Biopersistence in the Lung. *ACS Nano* **2016**, *10*, 8054–8066. [CrossRef] [PubMed]
5. van Kesteren, P.C.; Cubadda, F.; Bouwmeester, H.; van Eijkeren, J.C.; Dekkers, S.; de Jong, W.H.; Oomen, A.G. Novel insights into the risk assessment of the nanomaterial synthetic amorphous silica, additive E551, in food. *Nanotoxicology* **2015**, *9*, 442–452. [CrossRef] [PubMed]
6. Fruijtier-Polloth, C. The safety of nanostructured synthetic amorphous silica (SAS) as a food additive (E 551). *Arch. Toxicol.* **2016**, *90*, 2885–2916. [CrossRef] [PubMed]
7. Murugadoss, S.; Lison, D.; Godderis, L.; Van Den Brule, S.; Mast, J.; Brassinne, F.; Sebaihi, N.; Hoet, P.H. Toxicology of silica nanoparticles: An update. *Arch. Toxicol.* **2017**, *91*, 2967–3010. [CrossRef]
8. Yazdimamaghani, M.; Moos, P.J.; Dobrovolskaia, M.A.; Ghandehari, H. Genotoxicity of amorphous silica nanoparticles: Status and prospects. *Nanomedicine Nanotechnol. Boil. Med.* **2019**, *16*, 106–125. [CrossRef]
9. Asweto, C.O.; Hu, H.; Liang, S.; Wang, L.; Liu, M.; Yang, H.; Duan, J.; Sun, Z. Gene profiles to characterize the combined toxicity induced by low level co-exposure of silica nanoparticles and benzo[a]pyrene using whole genome microarrays in zebrafish embryos. *Ecotoxicol. Environ. Saf.* **2018**, *163*, 47–55. [CrossRef]
10. Asweto, C.O.; Wu, J.; Hu, H.; Feng, L.; Yang, X.; Duan, J.; Sun, Z. Combined Effect of Silica Nanoparticles and Benzo[a]pyrene on Cell Cycle Arrest Induction and Apoptosis in Human Umbilical Vein Endothelial Cells. *Int. J. Environ. Res. Public Health* **2017**, *14*, 289. [CrossRef]
11. Wu, J.; Shi, Y.; Asweto, C.O.; Feng, L.; Yang, X.; Zhang, Y.; Hu, H.; Duan, J.; Sun, Z. Co-exposure to amorphous silica nanoparticles and benzo[a]pyrene at low level in human bronchial epithelial BEAS-2B cells. *Environ. Sci. Pollut. Res. Int.* **2016**, *23*, 23134–23144. [CrossRef] [PubMed]
12. Wu, J.; Zhang, J.; Nie, J.; Duan, J.; Shi, Y.; Feng, L.; Yang, X.; An, Y.; Sun, Z. The chronic effect of amorphous silica nanoparticles and benzo[a]pyrene co-exposure at low dose in human bronchial epithelial BEAS-2B cells. *Toxicol. Res.* **2019**, *8*, 731–740. [CrossRef] [PubMed]
13. Lu, C.F.; Li, L.Z.; Zhou, W.; Zhao, J.; Wang, Y.M.; Peng, S.Q. Silica nanoparticles and lead acetate co-exposure triggered synergistic cytotoxicity in A549 cells through potentiation of mitochondria-dependent apoptosis induction. *Environ. Toxicol. Pharmacol.* **2017**, *52*, 114–120. [CrossRef] [PubMed]
14. Ahamed, M.; Akhtar, M.J.; Alhadlaq, H.A. Co-Exposure to SiO_2 Nanoparticles and Arsenic Induced Augmentation of Oxidative Stress and Mitochondria-Dependent Apoptosis in Human Cells. *Int. J. Environ. Res. Public Health* **2019**, *16*, 3199. [CrossRef] [PubMed]
15. Cao, X.Q.; DeLoid, G.M.; Bitounis, D.; De La Torre-Roche, R.; White, J.C.; Zhang, Z.Y.; Ho, C.G.; Ng, K.W.; Eitzer, B.D.; Demokritou, P. Co-exposure to the food additives SiO_2 (E551) or TiO_2 (E171) and the pesticide boscalid increases cytotoxicity and bioavailability of the pesticide in a tri-culture small intestinal epithelium model: Potential health implications. *Environ. Sci.-Nano* **2019**, *6*, 2786–2800. [CrossRef]
16. Costantini, L.M.; Gilberti, R.M.; Knecht, D.A. The phagocytosis and toxicity of amorphous silica. *PLoS ONE* **2011**, *6*, e14647. [CrossRef]
17. Dalzon, B.; Aude-Garcia, C.; Collin-Faure, V.; Diemer, H.; Béal, D.; Dussert, F.; Fenel, D.; Schoehn, G.; Cianférani, S.; Carrière, M.; et al. Differential proteomics highlights macrophage-specific responses to amorphous silica nanoparticles. *Nanoscale* **2017**, *9*, 9641–9658. [CrossRef]
18. Kim, B.; Kim, H.; Yu, I.J. Assessment of nanoparticle exposure in nanosilica handling process: Including characteristics of nanoparticles leaking from a vacuum cleaner. *Ind. Health* **2014**, *52*, 152–162. [CrossRef]
19. Oh, S.; Kim, B.; Kim, H. Comparison of nanoparticle exposures between fumed and sol-gel nano-silica manufacturing facilities. *Ind. Health* **2014**, *52*, 190–198. [CrossRef]
20. Tarantini, A.; Douki, T.; Personnaz, M.B.; Besombes, J.L.; Jafrezzo, J.L.; Maitre, A. Effect of the chemical composition of organic extracts from environmental and industrial atmospheric samples on the genotoxicity of polycyclic aromatic hydrocarbons mixtures. *Toxicol. Environ. Chem.* **2011**, *93*, 941–954. [CrossRef]
21. Tarantini, A.; Maitre, A.; Lefebvre, E.; Marques, M.; Marie, C.; Ravanat, J.L.; Douki, T. Relative contribution of DNA strand breaks and DNA adducts to the genotoxicity of benzo[a]pyrene as a pure compound and in complex mixtures. *Mutat. Res.* **2009**, *671*, 67–75. [CrossRef] [PubMed]

22. Nikolova, T.; Marini, F.; Kaina, B. Genotoxicity testing: Comparison of the γH2AX focus assay with the alkaline and neutral comet assays. *Mutat. Res.* **2017**, *822*, 10–18. [CrossRef] [PubMed]
23. Lesuffleur, T.; Porchet, N.; Aubert, J.P.; Swallow, D.; Gum, J.R.; Kim, Y.S.; Real, F.X.; Zweibaum, A. Differential expression of the human mucin genes MUC1 to MUC5 in relation to growth and differentiation of different mucus-secreting HT-29 cell subpopulations. *J. Cell Sci.* **1993**, *106*, 771–783. [PubMed]
24. Dorier, M.; Tisseyre, C.; Dussert, F.; Béal, D.; Arnal, M.E.; Douki, T.; Valdiglesias, V.; Laffon, B.; Fraga, S.; Brandão, F.; et al. Toxicological impact of acute exposure to E171 food additive and TiO_2 nanoparticles on a co-culture of Caco-2 and HT29-MTX intestinal cells. *Mutat. Res.* **2019**, *845*, 402980. [CrossRef] [PubMed]
25. Ferraro, D.; Anselmi-Tamburini, U.; Tredici, I.G.; Ricci, V.; Sommi, P. Overestimation of nanoparticles-induced DNA damage determined by the comet assay. *Nanotoxicology* **2016**, *10*, 861–870. [CrossRef]
26. Magdolenova, Z.; Lorenzo, Y.; Collins, A.; Dusinska, M. Can standard genotoxicity tests be applied to nanoparticles? *J. Toxicol. Environ. Health Part A* **2012**, *75*, 800–806. [CrossRef] [PubMed]
27. Ravanat, J.L.; Duretz, B.; Guiller, A.; Douki, T.; Cadet, J. Isotope dilution high-performance liquid chromatography-electrospray tandem mass spectrometry assay for the measurement of 8-oxo-7,8-dihydro-2′-deoxyguanosine in biological samples. *J. Chromatogr. B Biomed. Sci. Appl.* **1998**, *715*, 349–356. [CrossRef]
28. Pfaffl, M.W.; Tichopad, A.; Prgomet, C.; Neuvians, T.P. Determination of stable housekeeping genes, differentially regulated target genes and sample integrity: BestKeeper—Excel-based tool using pair-wise correlations. *Biotechnol. Lett.* **2004**, *26*, 509–515. [CrossRef]
29. Pfaffl, M.W. A new mathematical model for relative quantification in real-time RT-PCR. *Nucleic Acids Res.* **2001**, *29*, e45. [CrossRef]
30. Drasler, B.; Sayre, P.; Steinhauser, K.G.; Petri-Fink, A.; Rothen-Rutishauser, B. In vitro approaches to assess the hazard of nanomaterials (vol 8, pg 99, 2017). *Nanoimpact* **2018**, *9*, 51. [CrossRef]
31. Genies, C.; Maître, A.; Lefèbvre, E.; Jullien, A.; Chopard-Lallier, M.; Douki, T. The extreme variety of genotoxic response to benzo[a]pyrene in three different human cell lines from three different organs. *PLoS ONE* **2013**, *8*, e78356. [CrossRef] [PubMed]
32. Lindahl, T.; Wood, R.D. Quality control by DNA repair. *Science* **1999**, *286*, 1897–1905. [CrossRef] [PubMed]
33. Wyatt, M.D.; Pittman, D.L. Methylating agents and DNA repair responses: Methylated bases and sources of strand breaks. *Chem. Res. Toxicol.* **2006**, *19*, 1580–1594. [CrossRef] [PubMed]
34. Champion, J.A.; Walker, A.; Mitragotri, S. Role of particle size in phagocytosis of polymeric microspheres. *Pharm. Res.* **2008**, *25*, 1815–1821. [CrossRef] [PubMed]
35. Oh, N.; Park, J.H. Endocytosis and exocytosis of nanoparticles in mammalian cells. *Int. J. Nanomed.* **2014**, *9* (Suppl. S1), 51–63. [CrossRef]
36. Dorier, M.; Beal, D.; Tisseyre, C.; Marie-Desvergne, C.; Dubosson, M.; Barreau, F.; Houdeau, E.; Herlin-Boime, N.; Rabilloud, T.; Carriere, M. The food additive E171 and titanium dioxide nanoparticles indirectly alter the homeostasis of human intestinal epithelial cells in vitro. *Environ. Sci.-Nano* **2019**, *6*, 1549–1561. [CrossRef]
37. Di Cristo, L.; Movia, D.; Bianchi, M.G.; Allegri, M.; Mohamed, B.M.; Bell, A.P.; Moore, C.; Pinelli, S.; Rasmussen, K.; Riego-Sintes, J.; et al. Proinflammatory Effects of Pyrogenic and Precipitated Amorphous Silica Nanoparticles in Innate Immunity Cells. *Toxicol. Sci.* **2016**, *150*, 40–53. [CrossRef]
38. Zhang, H.; Dunphy, D.R.; Jiang, X.; Meng, H.; Sun, B.; Tarn, D.; Xue, M.; Wang, X.; Lin, S.; Ji, Z.; et al. Processing pathway dependence of amorphous silica nanoparticle toxicity: Colloidal vs pyrolytic. *J. Am. Chem. Soc.* **2012**, *134*, 15790–15804. [CrossRef]
39. Klein, G.; Devineau, S.; Aude, J.C.; Boulard, Y.; Pasquier, H.; Labarre, J.; Pin, S.; Renault, J.P. Interferences of Silica Nanoparticles in Green Fluorescent Protein Folding Processes. *Langmuir* **2016**, *32*, 195–202. [CrossRef]
40. Feng, L.; Yang, X.; Shi, Y.; Liang, S.; Zhao, T.; Duan, J.; Sun, Z. Co-exposure subacute toxicity of silica nanoparticles and lead acetate on cardiovascular system. *Int. J. Nanomed.* **2018**, *13*, 7819–7834. [CrossRef]
41. Guo, M.; Xu, X.; Yan, X.; Wang, S.; Gao, S.; Zhu, S. In vivo biodistribution and synergistic toxicity of silica nanoparticles and cadmium chloride in mice. *J. Hazard. Mater.* **2013**, *260*, 780–788. [CrossRef] [PubMed]
42. Hu, H.; Shi, Y.; Zhang, Y.; Wu, J.; Asweto, C.O.; Feng, L.; Yang, X.; Duan, J.; Sun, Z. Comprehensive gene and microRNA expression profiling on cardiovascular system in zebrafish co-exposured of SiNPs and MeHg. *Sci. Total Environ.* **2017**, *607*, 795–805. [CrossRef] [PubMed]

43. Yang, X.; Feng, L.; Zhang, Y.; Hu, H.; Shi, Y.; Liang, S.; Zhao, T.; Cao, L.; Duan, J.; Sun, Z. Co-exposure of silica nanoparticles and methylmercury induced cardiac toxicity in vitro and in vivo. *Sci. Total Environ.* **2018**, *631*, 811–821. [CrossRef] [PubMed]
44. Yu, Y.; Duan, J.; Li, Y.; Yu, Y.; Jin, M.; Li, C.; Wang, Y.; Sun, Z. Combined toxicity of amorphous silica nanoparticles and methylmercury to human lung epithelial cells. *Ecotoxicol. Environ. Saf.* **2015**, *112*, 144–152. [CrossRef] [PubMed]
45. Nakamura, M.; Imaoka, S.; Amano, F.; Funae, Y. P450 isoforms in a murine macrophage cell line, RAW264.7, and changes in the levels of P450 isoforms by treatment of cells with lipopolysaccharide and interferon-gamma. *Biochim. Biophys. Acta* **1998**, *1385*, 101–106. [CrossRef]
46. Torres, A.; Dalzon, B.; Collin-Faure, V.; Rabilloud, T. Repeated vs. Acute Exposure of RAW264.7 Mouse Macrophages to Silica Nanoparticles: A Bioaccumulation and Functional Change Study. *Nanomaterials* **2020**, *10*, 215. [CrossRef]
47. Demir, E.; Castranova, V. Genotoxic effects of synthetic amorphous silica nanoparticles in the mouse lymphoma assay. *Toxicol. Rep.* **2016**, *3*, 807–815. [CrossRef]
48. Haase, A.; Dommershausen, N.; Schulz, M.; Landsiedel, R.; Reichardt, P.; Krause, B.C.; Tentschert, J.; Luch, A. Genotoxicity testing of different surface-functionalized SiO_2, ZrO_2 and silver nanomaterials in 3D human bronchial models. *Arch. Toxicol.* **2017**, *91*, 3991–4007. [CrossRef]
49. Maser, E.; Schulz, M.; Sauer, U.G.; Wiemann, M.; Ma-Hock, L.; Wohlleben, W.; Hartwig, A.; Landsiedel, R. In vitro and in vivo genotoxicity investigations of differently sized amorphous SiO_2 nanomaterials. *Mutat. Res. Genet. Toxicol. Environ. Mutagen.* **2015**, *794*, 57–74. [CrossRef]
50. Wills, J.W.; Hondow, N.; Thomas, A.D.; Chapman, K.E.; Fish, D.; Maffeis, T.G.; Penny, M.W.; Brown, R.A.; Jenkins, G.J.; Brown, A.P.; et al. Genetic toxicity assessment of engineered nanoparticles using a 3D in vitro skin model (EpiDerm™). *Part. Fibre Toxicol.* **2016**, *13*, 50. [CrossRef]

Article

Silica Nanoparticles Provoke Cell Death Independent of p53 and BAX in Human Colon Cancer Cells

Susanne Fritsch-Decker, Zhen An, Jin Yan, Iris Hansjosten, Marco Al-Rawi, Ravindra Peravali, Silvia Diabaté * and Carsten Weiss *

Institute of Toxicology and Genetics, Karlsruhe Institute of Technology, Hermann-von-Helmholtz-Platz 1, 76344 Eggenstein-Leopoldshafen, Germany

* Correspondence: silvia.diabate@kit.edu (S.D.); carsten.weiss@kit.edu (C.W.); Tel.: +49-72160822692 (S.D.); +49-72160824906 (C.W.)

Received: 10 July 2019; Accepted: 12 August 2019; Published: 16 August 2019

Abstract: Several in vitro studies have suggested that silica nanoparticles (NPs) might induce adverse effects in gut cells. Here, we used the human colon cancer epithelial cell line HCT116 to study the potential cytotoxic effects of ingested silica NPs in the presence or absence of serum. Furthermore, we evaluated different physico-chemical parameters important for the assessment of nanoparticle safety, including primary particle size (12, 70, 200, and 500 nm) and surface modification (–NH_2 and –COOH). Silica NPs triggered cytotoxicity, as evidenced by reduced metabolism and enhanced membrane leakage. Automated microscopy revealed that the silica NPs promoted apoptosis and necrosis proportional to the administered specific surface area dose. Cytotoxicity of silica NPs was suppressed by increasing amount of serum and surface modification. Furthermore, inhibition of caspases partially prevented silica NP-induced cytotoxicity. In order to investigate the role of specific cell death pathways in more detail, we used isogenic derivatives of HCT116 cells which lack the pro-apoptotic proteins p53 or BAX. In contrast to the anticancer drug cisplatin, silica NPs induced cell death independent of the p53–BAX axis. In conclusion, silica NPs initiated cell death in colon cancer cells dependent on the specific surface area and presence of serum. Further studies in vivo are warranted to address potential cytotoxic actions in the gut epithelium. The unintended toxicity of silica NPs as observed here could also be beneficial. As loss of p53 in colon cancer cells contributes to resistance against anticancer drugs, and thus to reoccurrence of colon cancer, targeted delivery of silica NPs could be envisioned to also deplete p53 deficient tumor cells.

Keywords: synthetic amorphous silica; nanoparticles; colon cells; in vitro toxicity; cell death

1. Introduction

Synthetic amorphous silica (SAS) nanoparticles (NPs) are produced in large amounts for applications in industry and medicine. Many consumer products including food contain SAS, which occurs in different forms depending on the process of its manufacture. Food-grade SAS (E551) includes fumed (pyrogenic) silica and hydrated silica (precipitated silica, silica gel, and hydrous silica). Colloidal silica (silica sol) is not authorized as a food additive [1], but is an emerging material for various biomedical applications such as drug delivery [2]. Addition of SAS to food products has, for example, stabilizing, anti-caking, anti-settling, and emulsifying effects [3].

SAS food additive has been used for decades and is considered to be safe for consumers. Yang et al. [4] reported the occurrence of SAS at 1.3–16.3 mg Si/g dry food product, mainly in processed food such as coffee creamer, pudding powder, cake mix, or in probiotic tablets. In all exposure scenarios studied by EFSA (European Food Safety Authority), the lowest exposure was reported in the elderly, while the highest was in infants and in children, ranging from 3.9 to 74.2 mg/kg body weight per day and from 8.4 to 162.7 mg/kg body weight per day, respectively [1].

The effects of oral uptake of food-grade SAS nanoparticles have been reviewed recently [1,3,5–8]. In vivo studies with rats found no acute toxicity due to SAS ingestion, but fibrosis in the liver at high dosages [9]. In a study with mice, upon oral exposure to SAS NPs, increased pro-inflammatory cytokine levels (IL-1β, IL-6, and TNF-α) were detected in the colons. Additionally, ingested SAS NPs increased the richness and diversity of microbial species within the intestinal tract [10]. Recently, Siemer et al. found that nanomaterials including SAS impact the (patho)biology of bacteria occurring in the gut [11].

In addition, in vitro studies suggest that food-grade SAS is potentially hazardous. In particular, fumed SAS is more toxic than previously assumed. The study of Zhang et al. [12] showed that fumed silica was more active in producing ROS (Reactive Oxygen Species) and causing red blood cell hemolysis. In Caco-2 cells, fumed SiO_2 NPs induced cytotoxicity, DNA damage, and glutathione (GSH) depletion in serum-free medium [13,14]. Using human HT29 colon cells, it has been demonstrated that silica NPs are taken up into cells and stimulate cell proliferation under high-serum conditions. Under low-serum-conditions, cytotoxic effects were observed [15]. The study also found that SiO_2-NPs enhanced the biosynthesis of GSH via the mitogen activated protein kinase (MAPK) pathway and caused oxidative DNA damage. Activation of the MAPK pathway and cytotoxicity in response to silica NPs were also observed in the human stomach cell line GXF252L [16].

Winter et al. found that SiO_2-NPs affected the viability of murine dendritic cells and activated the inflammasome, suggesting that oral administration of these NPs could promote intestinal inflammatory responses [17]. More recently, the Toll-like receptor signaling pathway was demonstrated to be upstream of the inflammasome, and is essential in murine dendritic cells for the induction and release of IL-1β in response to food-grade nanosilica [18].

Before the particles reach the cells of the intestinal mucosa, they must pass the gastrointestinal tract, encompassing the oral cavity, esophagus, stomach, and intestine, and are exposed to different chemical conditions. The fluids in the gastrointestinal lumen contain complex mixtures of biomolecules such as digestive enzymes and food in different stages of digestion. Furthermore, osmotic concentration, pH, and the gut microbiome change during the passage through the gastrointestinal tract. All these parameters affect the ingested NPs and can alter the physico-chemical properties of the particles, which may influence the toxicological outcome [19]. Therefore, it is very difficult or nearly impossible to mimic all the parameters and physiological conditions encountered in vivo in an in vitro experiment. Furthermore, it has been demonstrated that biomolecules, in particular, proteins, bind to the particle surface and form a so-called biomolecular corona [20,21], which has consequences for the biological activity of nanomaterials [22,23]

The epithelial barrier of the gut represents a target for potentially cytotoxic effects of ingested silica NPs. As in the absence of a protective protein corona silica NPs induce cell death, for example, in lung epithelial and phagocytic cells, reminiscent of apoptosis [24,25], we wanted to investigate the role of specific cell death pathways in intestinal epithelial cells in more detail. Therefore, we used the human colon epithelial cancer cell line HCT116 as a model to evaluate the cytotoxic effects of SiO_2-NPs as a function of different physico-chemical parameters important for the assessment of nanoparticle safety [26,27], i.e., concentration (10, 25, 50, and 100 μg/mL), duration of exposure (5, 24, and 48 h), particle size (12, 70, 200, and 500 nm), surface modification (–NH_2 and –COOH), and protein coating with fetal bovine serum (FBS). Besides the wild-type cells, we used their isogenic derivatives, which lack the p53 or bax gene [28], to clarify the role of these pro-apoptotic factors for SiO_2-NP-induced cell death.

2. Materials and Methods

2.1. Materials

Materials and reagents were obtained from the following suppliers: Dulbecco's modified Eagle's medium (DMEM) (cat no 41966), cell culture medium supplements, Hank's balanced salt solution (HBSS, cat no 14025), Dulbecco's phosphate-buffered saline (DPBS, cat no 14190094), BCA (bicinchoninic acid) protein quantitative assay kit (cat no 23225): ThermoFisher Scientific (Dreieich, Germany); fetal bovine serum (FBS),Lactate dehydrogenase (LDH)cytotoxicity detection kit (cat no 11644793001), Hoechst 33258 (Hoechst, cat no B2261), propidium iodide (PI, cat no P4170), CDDP [cisplatin, *cis*-diamineplatinum (II) dichloride, cat no P4394], chemicals for sodium dodecylsulfate polyacrylamide gel electrophoresis (SDS-PAGE), and standard laboratory chemicals: Sigma-Aldrich (Taufkirchen, Germany); AlamarBlue reagent: AbD Serotec (Puchheim, Germany, cat no BUF012B); Aerosil® 200 (SiO_2—12 nm): Evonik Industries (Frankfurt am Main, Germany); Stöber-synthesized silica SiO_2—70 nm cat no 43-00-701, SiO_2-NH_2—70 nm cat no 43-01-701, SiO_2-COOH—70 nm cat no 43-02-701, SiO_2—200 nm cat no 43-00-202, SiO_2—500 nm cat no 43-00-502: Micromod Partikeltechnologie (Rostock, Germany). All particles are listed in Table 1.

Enhanced chemiluminescence (ECL) reagents were obtained from GE Healthcare Amersham (cat no 2232). Antibodies for detection of cleaved caspase 3 (Asp 175, cat no 9661), caspase 8 (1C12, cat no 9746), and cleaved caspase 9 (Asp 315, cat no 9505) were purchased from Cell Signaling Technology (Frankfurt am Main, Germany). Antibodies against p53 (DO-1, cat no sc-126) and PCNA (PC-10, cat no sc-56) were from Santa Cruz (Heidelberg, Germany). Secondary horseradish-peroxidase (HRP)-conjugated antibodies: DAKO (Hamburg, Germany), pan-caspase inhibitor Q-VD-OPh: 5-(2,6-Difluorophenoxy)-3-[[3-methyl-1-oxo-2-[(2-quinolinylcarbonyl)amino]butyl]amino]-4-oxo-pentanoic acid hydrate: MP Biomedicals, Heidelberg, Germany, cat no 03OPH10901.

2.2. Particle Suspensions and Characterization

For treatment of cells, the particle stock suspensions were generated just before preparing the diluted particle suspensions in cell culture medium with or without serum, which were added to the cells. Aerosil® 200 NPs delivered as powder were suspended in sterile deionized water at 1 mg/mL, shortly vortexed and probe sonified with 15 strokes, 15% cycle duty, output control 5 (Branson Sonifier 250, Schwäbisch Gmünd, Germany). The other colloidal silica particles, which were delivered as suspension, were diluted to 1 mg/mL in deionized water and vortexed. These stock solutions were further diluted in medium to the desired concentrations.

For analysis by dynamic light scattering (DLS), NPs were further diluted to 50 µg/mL in deionized water or DMEM with or without serum. The samples were then analyzed directly after vortexing using the Zetasizer Nano ZS (Malvern Instruments Ldt., Herrenberg, Germany) at 25 °C.

For analysis by transmission electron microscopy (TEM), the particle suspensions in water were transferred onto TEM grids (Plano, SF162-6), dried, and analyzed by a Zeiss electron microscope (Zeiss 109T, Oberkochen, Germany) [29].

2.3. Calculation of the Effective Density and the Deposited Dose

To compare the effects of differently sized SiO_2 particles, it was necessary to calculate the real dose deposited onto the cells over 24 h, as already described previously [30]. Briefly, the effective density of the particles in the respective medium was determined by the volumetric centrifugation method (VCM) according to Deloid et al. [31] at the highest applied concentration of 100 µg/mL. 1 mL of the particle suspension was filled into a packed cell volume (PCV) tube (TPP Techno Plastic Products, Trasadingen, Switzerland) and centrifuged in a swinging bucket rotor at 3.000 *g* for 1 h to collect the agglomerates in the capillary section of the tube. The volume of the pellet was determined using a measuring device from TPP Techno Plastic Products (Trasadingen, Swizerland). The effective density was then calculated according to the formula given in Deloid et al. [32]. The relative in vitro dose (RID) was determined by calculating the particle mass deposited onto the cell surface after 24 h using the distorted grid (DG) nanotransport simulator [33], based on hydrodynamic size (Table S1), effective density (Table S2), and other parameters in the respective media.

2.4. Cells

HCT116 wt, p53$^{-/-}$, and BAX$^{-/-}$ cells (kindly provided by B. Vogelstein, John Hopkins University, Baltimore, MD, USA) were cultured as described before [34]. Briefly, the cells were maintained in Dulbecco's modified Eagle's medium (DMEM) supplemented with 10% fetal bovine serum (FBS), 100 U/mL penicillin and 100 mg/mL streptomycin at 37 °C in a humidified atmosphere containing 5% CO_2. Medium was changed every 2 days. Control cells were treated with medium alone or with 0.1% DMSO, the solvent used for the caspase inhibitor.

2.5. Cell Death Analysis by Fluorescence Microscopy

For detection of cell number and stages of cell death by automated microscopy, 8000 cells were seeded per well of a 96 well plate. On the next day, the cell culture medium was discarded and the cells were treated according to the experimental design. After the incubation period, analysis was performed as previously described [30]. Briefly, Hoechst 33342 and propidium iodide (PI) were added to a final concentration of 0.3 μg/mL and 0.5 μg/mL, respectively. After 30 min incubation in the dark, bright field (BF) and fluorescence images were acquired from four positions in the well using an automated Olympus IX81 fluorescence microscope and a 10× objective (Olympus, Hamburg, Germany). The Hoechst dye was detected at excitation and emission wavelengths of 350 and 450 nm, respectively. PI dye was detected at 488 nm and 590 nm, respectively. The images were analyzed by the scan^R analysis software (version 2.7.3, Olympus, Hamburg, Germany) to obtain the total number of cells (Hoechst channel) and the number of early apoptotic, late apoptotic, and necrotic cells (combination of Hoechst and PI channel), as described previously [30].

2.6. Real-Time Imaging at the Single Cell Level

For real-time imaging at single cell level, HCT116 cells were first seeded in 96 well plates, as described above, and incubated overnight. Real-time imaging was performed as published previously [25]. Briefly, before treatment with particles, cells were stained with 0.1 μg/mL Hoechst and 0.083 μg/mL PI for 1 h at 37 °C and 5% CO_2, followed by an incubation with SiO_2 NPs over 24 h in a microscope incubator box (EMBLEM, Heidelberg, Germany) under control of CO_2, humidity, and temperature (37 °C, 5% CO_2). Two images per well and channel (bright-field, Hoechst and PI) were acquired using the automated fluorescence microscope IX81 (Olympus, Hamburg, Germany) with a 20-fold objective. The NIH ImageJ Software (version 1.50b, Bethesda, MD, USA) was used to convert images to videos.

2.7. LDH Cytotoxicity Assay

The LDH assay was performed as described previously [35]. Briefly, after treatment with particles, control cells were treated with 1% (v/v) of Triton X-100 for 30 min to detect the maximum LDH release (positive control). The whole plates were then centrifuged at 1500 rpm for 5 min and 50 μL of the supernatant was transferred into a 96 well plate. 50 μL PBS and 100 μL of the LDH working solution, prepared according to the manufacturer's instructions, were added per well and incubated at room temperature for 10 min. By the addition of 50 μL of 1 N HCl, the reaction was stopped. Absorbance was measured at 490 nm using a multi-well plate reader and analyzed by the software package SoftMaxPro (version 3.0, Molecular Devices, Ismaning, Germany). Cytotoxicity data are depicted as percentage relative to the positive control set to 100%.

2.8. AlamarBlue Viability Assay

The AlamarBlue® assay was performed as described previously [35], using the same plate with exposed cells as for the LDH assay. The remaining medium was removed and 100 µL of diluted AlamarBlue® reagent (1:20 in HBSS) was added. After 60 min in the incubator, the samples were analyzed using a multi-well plate fluorescent reader (Bio-Tek FL600, software package KC4 version 2.7, MWG-Biotech AG, Ebersberg, Germany) at 560 nm excitation and 620 nm emission wavelengths. Viability data are depicted as percentage relative to the negative control (untreated cells, 100%).

2.9. Protein Detection by Western Blot

Western blot was performed as described before [36]. The cells, which were seeded into 6 well plates, were treated with the particles according to the experimental design. After treatment, the cells were lysed in 2× Laemmli buffer (125 mM Tris–HCl, 4% SDS, 20% glycerol, 8% beta-mercaptoethanol, pH 6.8), boiled at 95 °C for 5 min, and sonicated in an ultrasonic water bath (Bandelin Sonorex, Berlin, Germany) for 10 min. Lysates were loaded on 12 or 15% gels depending on the molecular size of the protein to be detected for SDS-PAGE. After electrophoresis, the proteins were transferred onto Immobilon membranes (Millipore, Darmstadt, Germany). Membrane blocking was performed in 5% (w/v) non-fat dry milk powder in Tris-buffered saline containing 1% (v/v) Tween20 (TBS-T) for 1 h. Appropriate primary antibodies were applied in 5% (w/v) non-fat dry milk in TBS-T overnight at 4 °C. After incubation with appropriate HRP-conjugated secondary antibodies for 1 h, enhanced chemiluminescence (ECL) detection was performed according to the manufacturers' instructions.

2.10. Statistical Analyses

Data are expressed as mean ± standard deviation (SD). The significance of difference between two mean values was assessed by the Student's *t*-test or the Mann–Whitney Rank Sum test using the SigmaPlot software (version 11.0, Systat Software GmbH, Erkrath, Germany). A p-value < 0.05 was considered to be statistically significant.

3. Results and Discussion

3.1. Characterization of Silica Particles

In this study, we used commercial amorphous silica nanoparticles synthesized by different methods and of different sizes and surface modifications, as listed in Table 1. SiO_2—12 nm, also known as Aerosil® 200, is a hydrophilic pyrogenic silica NP produced by flame synthesis and was delivered as a powder. SiO_2—70 nm NPs were produced by a wet process known as the Stöber method [37] and were delivered in an aqueous suspension. Amine- and carboxyl-modified silica NPs of the same size were used to study the effects of surface modification on the toxicity of colloidal silica NPs. Furthermore, the effect of size was studied by using 200 and 500 nm Stöber-synthesized silica particles. The primary particle diameter analyzed by TEM was used to calculate the specific surface area of the particles in m^2/g. The particles were further characterized in medium without and with 10% FBS by dynamic light scattering (Table S1). In DMEM without FBS, all Stöber-synthesized particles showed a hydrodynamic diameter very close to their primary diameter, indicating a monodisperse suspension. The hydrodynamic diameter was slightly increased in the medium with 10% FBS, most likely due to the formation of a protein corona. SiO_2—12 nm NPs were already aggregated to 232 nm in medium without FBS (presumably due to sintering [22]), and further agglomerated in medium with 10% FBS. A more detailed characterization of Aerosil® 200 according to OECD test guidelines is provided in Mülhopt et al. [38]. In these studies, trace amounts of Ni (0.1 µg/g) and Cu (0.2 µg/g) were detected by inductively coupled plasma-mass spectrometry (ICP-MS), which, however, were below the level of toxicological relevance.

Table 1. Characteristics of silica nano- and microparticles.

Particles	Surface Modification	Nominal Primary Particle Diameter (nm) [a]	Primary Particle Diameter (nm) [b]	Specific Surface Area (m^2/g)
SiO_2—12 nm (Aerosil® 200)	plain	12	15 ± 10 [d]	200 ± 25 [a]
SiO_2—70 nm	plain	70	55 ± 7 [e]	55 [c]
SiO_2-NH_2—70 nm	–NH_2	70	55 ± 7 [e]	55 [c]
SiO_2-COOH—70 nm	–COOH	70	64 ± 7	47 [c]
SiO_2—200 nm	plain	200	190 ± 20	16 [c]
SiO_2—500 nm	plain	500	433 ± 25	6.9 [c]

[a] Data provided by the supplier, [b] analyzed by transmission electron microscopy (TEM), [c] calculated from the primary particle size, and the density 2.0 g/cm^3 for SiO_2 nanoparticles (NPs). [d] Already published in Reference [38], [e] already published in Reference [25].

3.2. Silica NPs Induced Apoptotic and Necrotic Cell Death Which Was Suppressed by the Presence of Serum

The effects of SiO_2—12 nm NPs on cell viability after 24 h were tested using the AlamarBlue and the LDH release assays (Figure 1a,b). Viable cells reduce the AlamarBlue dye by mitochondrial dehydrogenases and thus change its color. Reduction of AlamarBlue did not change when cells were exposed to particles dispersed in medium with 10% FBS. However, it decreased dose-dependently after treatment in the absence of FBS, even at 10 µg/mL of SiO_2—12 nm NPs. The results of the LDH assay correlate well with the AlamarBlue assay. Release of LDH is an indicator of plasma membrane damage. For cells treated with 10 µg/mL of SiO_2—12 nm NPs in FBS-free medium, there was already a strong release of LDH, which further increased with higher doses. There was no LDH release detected when cells were exposed to particles suspended in medium with 10% FBS. The cytotoxic effects of SiO_2—12 nm NPs observed after 48 h were similar to those after exposure for 24 h (Figure S1a,b), indicating a rapid onset of cell damage. Indeed, silica NPs triggered membrane rupture as early as 5 h after exposure (Figure S2a). In order to mimic the exposure of colon cells from the luminal side, we further reduced the amount of serum in the exposure medium. Interestingly, silica NPs also provoked membrane damage at 24 h in the presence of 1% FBS, albeit reduced compared to the levels induced in the total absence of serum (Figure S2b). The protective effects of just 1% FBS in the exposure medium were even more obvious at an earlier time point, i.e., 5 h after treatment (Figure S2a). In conclusion, silica NPs promoted cytotoxicity in HCT116 cells dependent on time, concentration, and the presence of serum, as previously reported for other cell types [15,16,20,22,23,39].

Next, we used automated fluorescence microscopy to further identify the mechanisms of cell death. After 24 h in the absence of FBS, the total cell number decreased significantly at increasing levels of SiO_2—12 nm NPs, while, concomitantly, the proportion of dead cells increased (Figure 2c,d). Similar effects were found at 48 h of exposure (Figure S3). Dead cells showed typical features of cell death, such as DNA condensation and loss of membrane integrity, and were mainly categorized as late apoptotic and necrotic cells (Figure 2a,b,d, Video S1). As also documented above using the AlamarBlue and LDH assays (Figure 1), 10% serum in the exposure medium largely prevented cytotoxicity. In accordance with the rapid damage of the cell membrane detected by the LDH assay at 5 h in the absence of serum (Figure S2a), an early influx of propidium iodide indicative of membrane rupture was also observed at the single cell level (Figure S4a,b). Furthermore, in accordance with the LDH assay (Figure S2b), the presence of 1% FBS reduced cell death prompted by silica NPs (Figure S5a,b). These findings, together with previous results obtained with human colon carcinoma [15] and lung epithelial cells [24], indicate that the ratio of the specific particle surface area and concentration of proteins in the exposure medium has a critical impact on the occurrence of cell death upon treatment with silica NPs.

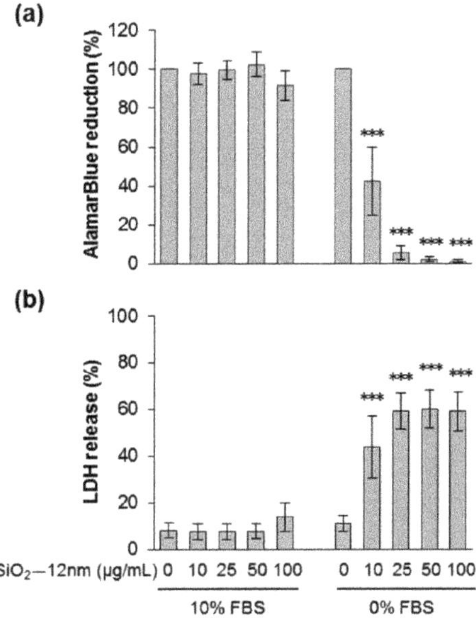

Figure 1. Cell viability decreased after incubation with silica NPs in the absence of fetal bovine serum (FBS). HCT116 wt cells were incubated with SiO_2—12 nm NPs at the indicated concentration in the presence or absence of 10% FBS. After 24 h, cell viability was detected by the AlamarBlue assay (**a**). The values were normalized to the negative control (no particles were added, 100%). Lactate dehydrogenase (LDH) release is shown in (**b**). The values were normalized to the positive control (1% Triton X-100, 100%). The data represent the means of seven independent experiments ± SD performed with six replicates. *** $p < 0.001$ indicates significant differences in the response of cells treated with the same amount of NPs in the absence (0% FBS) or the presence of serum (10% FBS).

3.3. Silica NPs Triggered Cell Death Dependent on Size, Specific Surface Area, and Surface Modification

To address the impact of size on silica NP induced cell death under serum-free conditions, we studied colloidal silica spheres with nominal diameters of 70, 200, and 500 nm (Table 1). Administration of particles with increasing diameter but at the same mass concentration (10 and 50 µg/mL) revealed an inverse relationship of particle size and cytotoxicity, i.e., larger particles were less toxic compared to smaller particles (Figure 3a). Similarly to the effects observed for SiO_2—12 nm NPs derived from flame synthesis (Figure 2), colloidal silica NPs also provoked apoptotic and necrotic cell death dependent on dose and size (Figure 3b, Figure S6). Silica NPs are surface active materials and interact with biological surfaces, i.e., cell membranes. The specific particle surface area (SSA, cm^2 at a defined mass or volume) increases linearly with a reciprocal decrease in particle size. The dose–response related to the nominal concentration in the exposure media indicated increased toxicity of smaller versus larger silica NPs (Figure 3a,b). However, normalization of the dose response curves to the administered SSA showed a clear correlation of SSA and cytotoxicity (Figure 3c). As the deposited, i.e., effective, dose of particles is controlled by sedimentation and diffusion, which depend on particle size and effective density, we also calculated the delivered cellular dose [33,40]. While for the larger particles of 200 and 500 nm there was no major difference in the effective SSA dose, for the smaller nanoparticles (12 and 70 nm), the effective SSA (Figure 3e,f) was drastically reduced compared to the nominal dose (Figure 3c,d). As deposition of nanoparticles is mainly driven by diffusion, the relative fraction of administered NPs interacting with cells was lower compared to the larger particles, highlighting the importance of computational modeling of the relative in vitro dose for hazard ranking of differently sized materials. Indeed, several in vitro and in vivo studies have shown that cytotoxicity of insoluble NPs better correlates with the SSA dose than with the particle mass dose [24,41–43].

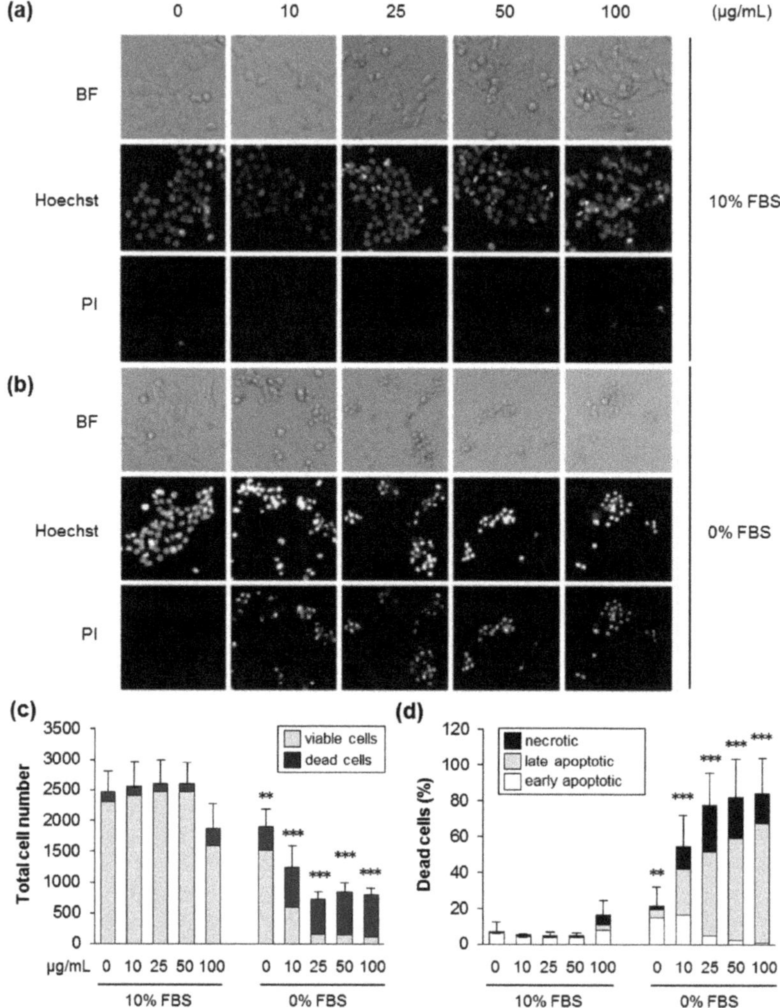

Figure 2. Silica NPs induce apoptotic and necrotic cell death in the absence of serum. HCT116 wt cells were incubated with SiO_2—12 nm NPs at the indicated concentration in the presence (10% FBS) or absence (0% FBS) of serum. After 24 h, the cells were stained with Hoechst and propidium iodide (PI), and images were acquired by automated microscopy and analyzed by the scan^R software to deduce cell numbers and the different modes of cell death. (**a**,**b**) Representative images of cells treated as indicated in the brightfield, the Hoechst, and the PI channels. (**c**) The total cell number divided into living and dead cells after treatment, as indicated. (**d**) The percentage of dead cells relative to the total cell number divided into the different classes of cell death, as indicated. Data are represented as mean values of three independent experiments carried out with four replicates (n = 12). The error bars are SD values related to the total cell number (**c**) or the percentage of dead cells (**d**). ** $p < 0.01$, *** $p < 0.001$ indicate significant differences in the response of cells treated with particles at corresponding concentrations in the absence (0% FBS) or the presence of serum (10% FBS).

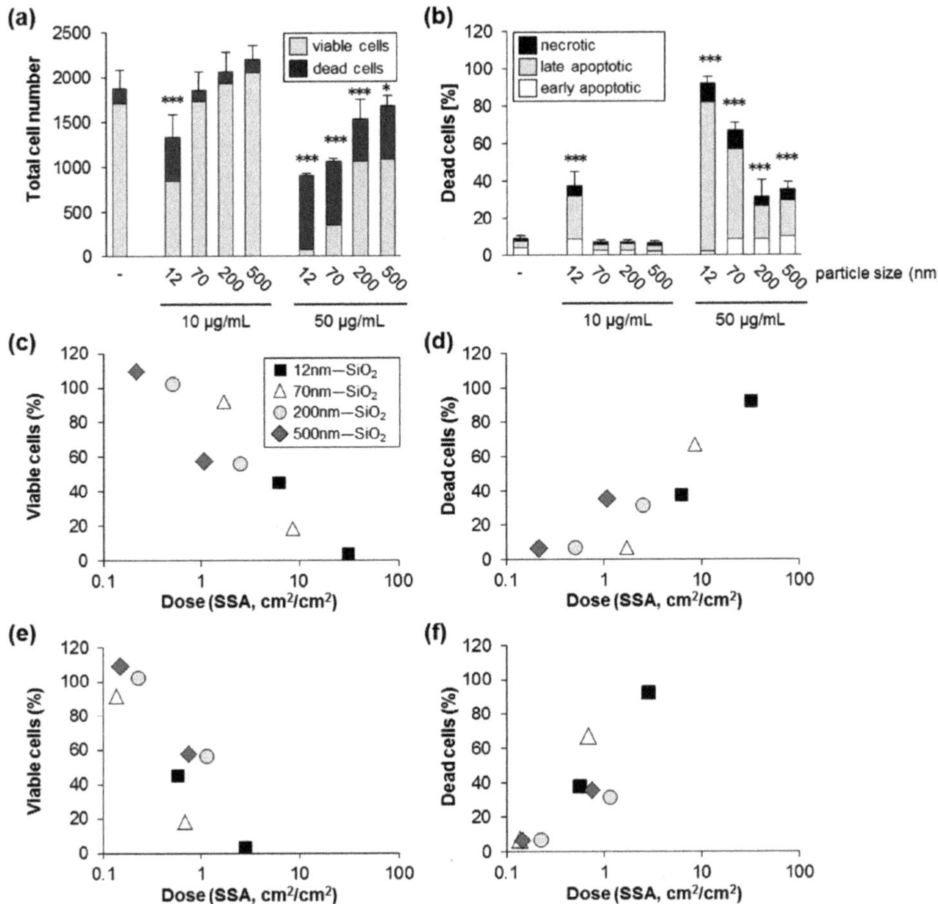

Figure 3. Silica NPs induce cell death dependent on size and specific surface area. HCT116 wt cells were incubated with particles as indicated in the absence of FBS. After 24 h, the cells were processed and analyzed as described in Figure 2. (**a**) The total cell number divided into living and dead cells. (**b**) The percentage of dead cells relative to the total cell number divided into the different classes of cell death as indicated. Data show mean values of two independent experiments carried out with four replicates ($n = 8$). The error bars are SD values related to the total cell number (**a**) or the percentage of total dead cells (**b**), respectively. * $p < 0.05$, *** $p < 0.001$ indicate significant differences in the response of cells treated with particles in comparison to untreated control cells (-). The percentage of viable (**c**) and dead cells (**d**) are plotted against the nominal specific particle surface area dose per cell area (cm^2/cm^2, logarithmic scale). For comparison (**e**,**f**), the calculated deposited dose has been used as a metric. Note that the nominal (**c**,**d**) and calculated (**e**,**f**) dose in case of larger particles (200 and 500 nm) were rather similar, whereas for nanoparticles (12 and 70 nm), only a small fraction was deposited. Further details are described under Methods.

Chemical surface functionalization of silica NPs modulates the cellular response [44,45]. Therefore, we also investigated in HCT116 cells the effects of colloidal silica NPs of the same size (70 nm) with a plain, NH_2—, or COOH—modified surface. As observed in other cell models, chemical modification of the silica surface suppressed cytotoxicity (Figure 4a,b; Figure S7). Replacement of reactive silanol groups by functional groups is supposed to prevent their detrimental interaction with cellular membranes [46,47].

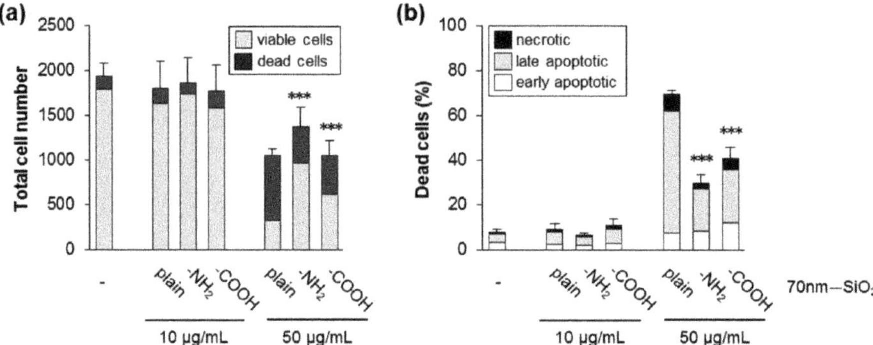

Figure 4. Surface modifications reduces cytotoxicity of silica NPs. HCT116 wt cells were exposed to SiO_2—NPs (70 nm) with a plain, NH_2—, or COOH—modified surface in the absence of FBS. After 24 h, the cells were processed and analyzed as described in Figure 2. (**a**) The total cell number divided into living and dead cells. (**b**) The percentage of dead cells relative to the total cell number divided into the different classes of cell death as indicated. Data show mean values of two independent experiments carried out with four replicates ($n = 8$). The error bars are SD values related to the total cell number (**a**) or the percentage of total dead cells (**b**), respectively. *** $p < 0.001$ indicates significant differences in the number of viable cells (**a**) and total dead cells (**b**) after administration of modified NPs compared to corresponding concentrations of plain particles.

3.4. Silica NPs Promoted Cell Death Dependent on Caspases and Independent of p53 or BAX

Silica NPs induced cell death in HCT116 cells via apoptosis and necrosis, as evidenced by the image analysis outlined above. Execution of cell death is controlled by intricate signaling networks and is of relevance for the toxicity of several nanomaterials [48]. The most prominent determinants of cell death are the tumor suppressor protein p53 and its target BAX [49]. p53 is a transcription factor which is activated due to cellular stress, and regulates cell growth, cell cycle progression, and cell death via apoptosis, necroptosis, or ferroptosis. BAX acts on mitochondria to promote apoptosis and is regulated by p53, which upregulates transcription of the gene encoding BAX. In order to further investigate the mechanism of cell death triggered by silica NPs, we explored wild-type HCT116 cells and their isogenic derivatives in which the p53 gene was deleted by homologous recombination [28]. Compared to wild-type cells, silica-NP-induced cell death monitored by automated microscopy at 24 h was not reduced in p53 knock-out cells (Figure 5a,b). The similar sensitivity of p53 deficient cells was also confirmed by the LDH and AlamarBlue assays (compare Figure S8a,b and Figure 1). Additionally, at an earlier time point (5 h), no reduced cell death but rather, enhanced cytotoxicity could be observed in p53 knock-out cells (Figure S9a,b; Figure S10; Figure S2a). Similarly, at 48 h after exposure to nanosilica, no major difference in the percentage of cell death was obvious when wildtype and p53 knock-out cells were compared (Figure S11). Next, we analyzed the involvement of the p53 target BAX in the execution of nanosilica-induced cell death. Again, as found in the case of p53 deficient cells, BAX knock-out cells were not protected against the detrimental action of nanosilica (Figure 5a,b). In summary and in contrast to the action of some genotoxins [28,34], the p53–BAX axis seems not to be involved in the execution of cell death prompted by exposure to nanosilica in HCT116 cells. Previous reports have suggested a pro-apoptotic role of p53 and BAX in nanosilica-induced cell death [50,51]. After treatment with silica NPs, an upregulation of the p53 and BAX proteins in human fetal hepatocytes and hepatoma cells could be demonstrated, which correlates with the onset of cell death. However, further studies in these and other cell types are warranted to really confirm a functional role of p53 and BAX (e.g., by genetic or pharmacological interference) in nanosilica-induced cell death.

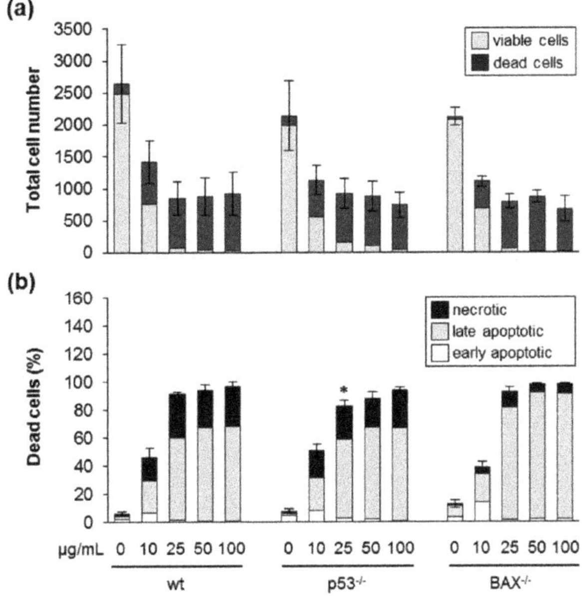

Figure 5. Silica NPs induce cell death independent of p53 and BAX. HCT116 wild-type (wt), p53 knock-out ($^{-/-}$) and BAX knock-out ($^{-/-}$) cells were incubated with SiO_2—12 nm NPs as indicated in the absence of FBS. After 24 h, the cells were processed and analyzed as described in Figure 2. (**a**) The total cell number divided into living and dead cells. (**b**) The percentage of dead cells relative to the total cell number divided into the different classes of cell death as indicated. Data show mean values of five–six independent experiments with wt and p53$^{-/-}$ cells. For BAX$^{-/-}$ cells, one representative experiment out of two is shown. All experiments were carried out with three–four replicates. The error bars are SD values related to the total cell number (**a**) or the percentage of total dead cells (**b**), respectively. Note that there was no major difference in the response of p53$^{-/-}$ or BAX$^{-/-}$ compared to wild-type cells. * $p < 0.05$.

Finally, we addressed the involvement of caspases in silica-NP-induced cell death. Indeed, inhibition of caspases partially protected cells from silica-NP-induced cell death, evidenced by an increase of viable cells (Figure 6a) and a decrease in apoptotic and necrotic cells (Figure 6b). As also shown for the positive controls, which were treated with the genotoxin cisplatin, caspases are required for the promotion of cell death initiated by silica NPs. Next, we monitored cleavage of caspase 8 and caspase 9, which act as initiator caspases in the extrinsic (as part of the death-inducing signaling complex) and intrinsic (as part of the apoptosome) apoptotic pathways, respectively [52]. Exposure of cells to increased amounts of silica NPs triggered cleavage of caspase 8 and 9 (Figure 6c), in line with a critical role of caspase activity in the execution of cell death (Figure 6a,b). Furthermore, in contrast to the positive control cisplatin, silica NPs did not elevate the protein levels of p53 (Figure 6c). This also supports the loss of function experiments, i.e., similar cytotoxicity in wt and p53 knock-out cells after exposure to silica NPs (Figure 5a,b).

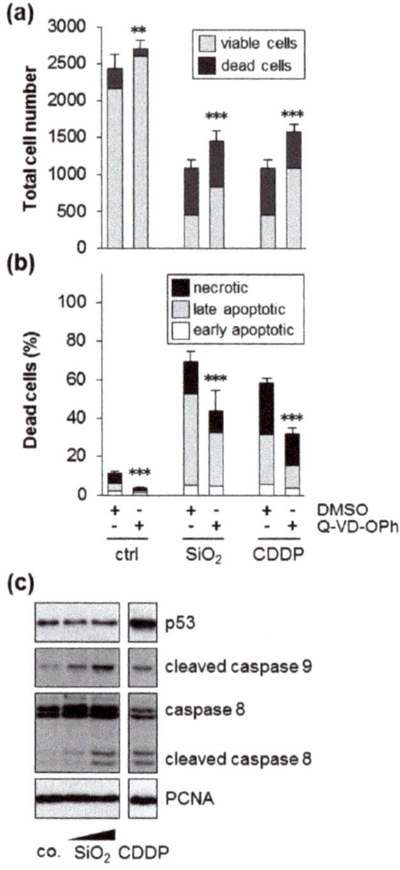

Figure 6. Silica NPs induce cell death dependent on caspases. HCT116 wt cells were incubated with 10 µg/mL SiO_2—12 nm NPs or with 50 µM of the positive control compound cisplatin (CDDP) in medium without FBS. Samples were co-treated as indicated either with the pan-caspase inhibitor Q-VD-OPh (10 µM) dissolved in DMSO (0.5%) or simply the solvent (DMSO, 0.5%). After 24 h, the cells were processed and analyzed as described in Figure 2. (**a**) The total cell number divided into living and dead cells; (**b**) the percentage of dead cells relative to the total cell number divided into the different classes of cell death as indicated. Data show mean values of two independent experiments carried out with four replicates ($n = 8$). The error bars are SD values related to the total cell number (**a**) or the percentage of total dead cells (**b**), respectively. *** $p < 0.001$ indicates significant differences in the response of cells treated with or without the pan-caspase inhibitor. (**c**) Activation, i.e., cleavage of caspase 8 and 9 was analyzed by western blotting after treatment with 0 (co.), 5 and 10 µg/mL SiO_2—12 nm NPs or with 50 µM CDDP in the absence of serum for 22 h. PCNA was used as a loading control. The results are representative of two independent experiments.

4. Conclusions and Perspectives

Engineered silica nanoparticles are ingested as food additives and belong, together with titania NPs, to the most frequently used nanomaterials in food products [8]. The present study revealed cytotoxic actions of silica NPs in the colon cancer cell line HCT116, specifically in the absence or presence of low amount of serum, corroborating previous studies in the two different human colon cancer cell lines Caco-2 and HT29 [13,14]. In Caco-2 cells, a pro-inflammatory action of silica NPs was also observed indicated by the release of IL-8 [53]. Interestingly, differentiated Caco-2 cells were less sensitive to the noxious effects of silica NPs. The physiological relevance of in vitro studies needs to be considered, and specific test systems and guidelines still need to be optimized in case of hazard and risk assessment of NPs upon ingestion [8]. With respect to dosimetry, we chose a realistic dose range for all different silica nanoparticles below the calculated relevant in vitro dose of about 2 $\mu g/cm^2$ (NP mass/cellular surface area), approximated from the daily average intake of food grade silica (1.8 mg/body weight/day) in humans [8]. Unfortunately, there is a lack of in vivo studies investigating toxicity of silica NPs in the gut. Recently, production of pro-inflammatory cytokines upon oral administration of nanosilica (2.5 mg/kg bw/day) for 7 days was observed in mice, which coincided with slight histological changes, i.e., crypt damage and inflammatory cell infiltration [10]. Thus, it is tempting to speculate that microlesions in the gut epithelium due to cell death could contribute to the inflammatory symptoms. In the colonic crypts, there is a constant renewal of epithelial cells, which originate from stem cells and progressively differentiate to become mature enterocytes. Clearly, more detailed studies in vivo are warranted to address the impact of nanosilica on the viability of epithelial cells, and also in relation to their differentiation status. It is of note that the in vitro experiments of Wiemann et al. with alveolar macrophages cultivated under serum-free conditions have been demonstrated to predict the short-term inhalation toxicity in vivo for 18 different nanomaterials, including nanosilica [54]. Upon inhalation, NPs are covered by the lung lining fluid, mostly comprised of surfactant (primarily phospholipids) and a small amount of proteins, and are not covered by serum proteins. Therefore, in vitro exposure of lung cells in the presence of high levels of serum proteins (i.e., 10% FBS) does not closely mimic the physiological conditions in the lung. Concerning a realistic exposure scenario for the gastrointestinal tract, NPs are embedded in a complex food matrix, which is further altered upon ingestion and digestion. Therefore, the corona, i.e., the adsorbed molecules on the NP surface, is ill-defined and comprised of a multitude of different molecules including carbohydrates, protein, and lipids. However, compared to localization of NPs in the blood stream after direct injection (in the case of medical applications) or after translocation from the gut into the portal vein, the amount of serum proteins in the gut lumen is much lower. Hence, the presence of high levels of serum proteins as used in most in vitro experiments might passivate the silica surface and suppress the interaction of silica NPs with critical targets such as the cellular membrane or certain receptors. Although in vitro experiments employing monocultures can be used to address specific processes and pathways, co-cultures of colon epithelial cells together with mucus producing goblet cells or lymphoid cells might even be better suited to predicting the more complex situation in vivo. Indeed, exposure of differentiated Caco-2 cells co-cultured with mucus producing HT29-MTX cells to titania NPs triggered the production of reactive oxygen species [55–57] and an inflammatory response, which was absent in Caco-2 cells grown as monocultures.

Finally, in vitro studies in toxicology often rely on the use of transformed cancer cells, as primary cells of various organs are difficult to cultivate and are often not available. Therefore, adverse effects observed in vitro might indicate toxicity in vivo, but need to be confirmed in follow-up investigations. However, the unintended toxicity of NPs determined in vitro might also be beneficial and exploited to selectively eliminate cancer cells. Specifically, silica NPs also kill p53-deficient colon cancer cells, as shown here, which are resistant to chemotherapy by conventional drugs such as cisplatin. Hence, oral delivery of silica NPs could be envisioned for the treatment of colon cancer. Selective accumulation of silica NPs via tumor-targeting ligands could be combined with the tunable unmasking of the silica surface, e.g., by dePEGylation initiated by near infrared radiation, as shown recently [58].

Supplementary Materials: The following are available online at http://www.mdpi.com/2079-4991/9/8/1172/s1, Table S1: Characterization of SiO_2 particle suspensions. Table S2: Nominal and effective i.e., calculated dose of differently sized SiO_2 particles. Figure S1: Cell viability decreases after incubation with SiO_2—12 nm NPs for 48 h in the absence of FBS. Figure S2: Low amount of serum (1% FBS) delays and reduces but does not totally prevent membrane damage provoked by SiO_2—12 nm NPs. Figure S3: Cell death upon treatment with SiO_2—12 nm NPs for 48 h in the absence of FBS. Figure S4: SiO_2—12 nm NPs disturb the integrity of the cell membrane in the absence of FBS as indicated by the influx of propidium iodide already after 5 h. Figure S5: The toxic effects of SiO_2—12 nm NPs after 24 h are diminished in the presence of 1% FBS. Figure S6: Silica NPs induce apoptotic and necrotic cell death dependent on size. Figure S7: Cytotoxicity of silica NPs is suppressed by surface modification. Figure S8: Cell viability decreases after incubation of HCT p53$^{-/-}$ cells with SiO_2—12 nm NPs for 24 h in the absence of FBS. Figure S9: HCT116 p53$^{-/-}$ cells are more sensitive to SiO_2—12 nm NPs than HCT116 wt cells. Figure S10: Low amount of serum (1% FBS) reduces membrane damage provoked by SiO_2—12 nm NPs. Figure S11: The response of HCT116 wt and p53$^{-/-}$ cells to SiO_2—12 nm NP exposure after 48 h is similar. Figure S12: Summary of the results from the different toxicity assays to demonstrate the similar sensitivity of HCT wt and p53$^{-/-}$ cells to SiO_2—12 nm NPs dependent on the concentration of FBS. Video S1: Time course of cell death after incubation to SiO_2—12 nm NPs in the absence of serum.

Author Contributions: Conceptualization, C.W.; methodology, I.H., M.A.-R., R.P.; investigation, Z.A., J.Y., S.F.-D.; writing—original draft preparation, S.D., C.W., S.F.-D.; writing—review and editing, S.D., C.W., S.F.-D.

Funding: Z.A. and J.Y. were funded by the China Scholarship Council (CSC).

Acknowledgments: We acknowledge support by the KIT-Publication Fund of the Karlsruhe Institute of Technology.

Conflicts of Interest: The authors declare no conflict of interest.

References

1. EFSA (European Food and Safety Authority). Re-evaluation of silicon dioxide (E551) as a food additive. *EFSA J.* **2018**, *16*, e05088.
2. Nyström, A.M.; Fadeel, B. Safety assessment of nanomaterials: Implications for nanomedicine. *J. Control. Release* **2012**, *161*, 403–408. [CrossRef] [PubMed]
3. Mebert, A.M.; Baglole, C.J.; Desimone, M.F.; Maysinger, D. Nanoengineered silica: Properties, applications and toxicity. *Food Chem. Toxicol.* **2017**, *109*, 753–770. [CrossRef] [PubMed]
4. Yang, Y.; Faust, J.J.; Schoepf, J.; Hristovski, K.; Capco, D.G.; Herckes, P.; Westerhoff, P. Survey of food-grade silica dioxide nanomaterial occurrence, characterization, human gut impacts and fate across its lifecycle. *Sci. Total Environ.* **2016**, *565*, 902–912. [CrossRef] [PubMed]
5. Winkler, H.C.; Suter, M.; Naegeli, H. Critical review of the safety assessment of nano-structured silica additives in food. *J. Nanobiotechnol.* **2016**, *14*, 44. [CrossRef] [PubMed]
6. Fruijtier-Pölloth, C. The safety of nanostructured synthetic amorphous silica (SAS) as a food additive (E 551). *Arch. Toxicol.* **2016**, *90*, 2885–2916. [CrossRef]
7. Murugadoss, S.; Lison, D.; Godderis, L.; van den Brule, S.; Mast, J.; Brassinne, F.; Sebaihi, N.; Hoet, P.H. Toxicology of silica nanoparticles: An update. *Arch.Toxicol.* **2017**, *91*, 2967–3010. [CrossRef]
8. Sohal, I.S.; O'Fallon, K.S.; Gaines, P.; Demokritou, P.; Bello, D. Ingested engineered nanomaterials: State of science in nanotoxicity testing and future research needs. *Part Fibre Toxicol.* **2018**, *15*, 29. [CrossRef] [PubMed]
9. Van der Zande, M.; Vandebriel, R.J.; Groot, M.J.; Kramer, E.; Herrera Rivera, Z.E.; Rasmussen, K.; Ossenkoppele, J.S.; Tromp, P.; Gremmer, E.R.; Peters, R.J.; et al. Sub-chronic toxicity study in rats orally exposed to nanostructured silica. *Part Fibre Toxicol.* **2014**, *11*, 8. [CrossRef]
10. Chen, H.; Zhao, R.; Wang, B.; Cai, C.; Zheng, L.; Wang, H.; Wang, M.; Ouyang, H.; Zhou, X.; Chai, Z.; et al. The effects of orally administered Ag, TiO_2 and SiO_2 nanoparticles on gut microbiota composition and colitis induction in mice. *Nanoimpact* **2017**, *8*, 80–88. [CrossRef]
11. Siemer, S.; Hahlbrock, A.; Vallet, C.; McClements, D.J.; Balszuweit, J.; Voskuhl, J.; Docter, D.; Wessler, S.; Knauer, S.K.; Westmeier, D.; et al. Nanosized food additives impact beneficial and pathogenic bacteria in the human gut: A simulated gastrointestinal study. *Sci. Food* **2018**, *22*, 1–10. [CrossRef] [PubMed]
12. Zhang, H.; Dunphy, D.R.; Jiang, X.; Meng, H.; Sun, B.; Tarn, D.; Xue, M.; Wang, X.; Lin, S.; Ji, Z.; et al. Processing pathway dependence of amorphous silica nanoparticle toxicity: Colloidal vs pyrolytic. *J. Am. Chem. Soc.* **2012**, *134*, 15790–15804. [CrossRef] [PubMed]
13. Gerloff, K.; Albrecht, C.; Boots, A.W.; Förster, I.; Schins, R.P.F. Cytotoxicity and oxidative DNA damage by nanoparticles in human intestinal Caco-2 cells. *Nanotoxicology* **2009**, *3*, 355–364. [CrossRef]
14. Gerloff, K.; Pereira, D.I.; Faria, N.; Boots, A.W.; Kolling, J.; Förster, I.; Albrecht, C.; Powell, J.J.; Schins, R.P. Influence of simulated gastrointestinal conditions on particle-induced cytotoxicity and interleukin-8 regulation in differentiated and undifferentiated Caco-2 cells. *Nanotoxicology* **2013**, *7*, 353–366. [CrossRef] [PubMed]
15. Gehrke, H.; Frühmesser, A.; Pelka, J.; Esselen, M.; Hecht, L.L.; Blank, H.; Schuchmann, H.P.; Gerthsen, D.; Marquardt, C.; Diabaté, S.; et al. In vitro toxicity of amorphous silica nanoparticles in human colon carcinoma cells. *Nanotoxicology* **2013**, *7*, 274–293. [CrossRef] [PubMed]

16. Wittig, A.; Gehrke, H.; Del, F.G.; Fritz, E.M.; Al-Rawi, M.; Diabaté, S.; Weiss, C.; Sami, H.; Ogris, M.; Marko, D. Amorphous silica particles relevant in food industry influence cellular growth and associated signaling pathways in human gastric carcinoma cells. *Nanomaterials* **2017**, *7*, 18. [CrossRef] [PubMed]
17. Winter, M.; Beer, H.D.; Hornung, V.; Kramer, U.; Schins, R.P.; Förster, I. Activation of the inflammasome by amorphous silica and TiO_2 nanoparticles in murine dendritic cells. *Nanotoxicology* **2011**, *5*, 326–340. [CrossRef] [PubMed]
18. Winkler, H.C.; Kornprobst, J.; Wick, P.; von Moos, L.M.; Trantakis, I.; Schraner, E.M.; Bathke, B.; Hochrein, H.; Suter, M.; Naegeli, H. MyD88-dependent pro-interleukin-1beta induction in dendritic cells exposed to food-grade synthetic amorphous silica. *Part Fibre Toxicol.* **2017**, *14*, 21. [CrossRef]
19. Bellmann, S.; Carlander, D.; Fasano, A.; Momcilovic, D.; Scimeca, J.A.; Waldman, W.J.; Gombau, L.; Tsytsikova, L.; Canady, R.; Pereira, D.I.; et al. Mammalian gastrointestinal tract parameters modulating the integrity, surface properties, and absorption of food-relevant nanomaterials. *Wiley Interdiscip. Rev. Nanomed. Nanobiotechnol.* **2015**, *7*, 609–622. [CrossRef]
20. Lesniak, A.; Fenaroli, F.; Monopoli, M.P.; Aberg, C.; Dawson, K.A.; Salvati, A. Effects of the presence or absence of a protein corona on silica nanoparticle uptake and impact on cells. *ACS Nano* **2012**, *6*, 5845–5857. [CrossRef]
21. Ruh, H.; Kühl, B.; Brenner-Weiss, G.; Hopf, C.; Diabaté, S.; Weiss, C. Identification of serum proteins bound to industrial nanomaterials. *Toxicol. Lett.* **2012**, *208*, 41–50. [CrossRef] [PubMed]
22. Panas, A.; Marquardt, C.; Nalcaci, O.; Bockhorn, H.; Baumann, W.; Paur, H.R.; Mülhopt, S.; Diabaté, S.; Weiss, C. Screening of different metal oxide nanoparticles reveals selective toxicity and inflammatory potential of silica nanoparticles in lung epithelial cells and macrophages. *Nanotoxicology* **2013**, *7*, 259–273. [CrossRef] [PubMed]
23. Docter, D.; Bantz, C.; Westmeier, D.; Galla, H.J.; Wang, Q.; Kirkpatrick, J.C.; Nielsen, P.; Maskos, M.; Stauber, R.H. The protein corona protects against size- and dose-dependent toxicity of amorphous silica nanoparticles. *Beilstein J. Nanotechnol.* **2014**, *5*, 1380–1392. [CrossRef] [PubMed]
24. Leibe, R.; Hsiao, I.L.; Fritsch-Decker, S.; Kielmeier, U.; Wagbo, A.M.; Voss, B.; Schmidt, A.; Hessman, S.D.; Duschl, A.; Oostingh, G.J.; et al. The protein corona suppresses the cytotoxic and pro-inflammatory response in lung epithelial cells and macrophages upon exposure to nanosilica. *Arch. Toxicol.* **2019**, *93*, 871–885. [CrossRef] [PubMed]
25. Hsiao, I.L.; Fritsch-Decker, S.; Leidner, A.; Al-Rawi, M.; Hug, V.; Diabaté, S.; Grage, S.L.; Meffert, M.; Stoeger, T.; Gerthsen, D.; et al. Biocompatibility of Amine-Functionalized Silica Nanoparticles: The Role of Surface Coverage. *Small* **2019**, *158*, e1805400. [CrossRef] [PubMed]
26. Gebel, T.; Foth, H.; Damm, G.; Freyberger, A.; Kramer, P.J.; Lilienblum, W.; Röhl, C.; Schupp, T.; Weiss, C.; Wollin, K.M.; et al. Manufactured nanomaterials: Categorization and approaches to hazard assessment. *Arch. Toxicol.* **2014**, *88*, 2191–2211. [CrossRef] [PubMed]
27. Lynch, I.; Weiss, C.; Valsami-Jones, E. A strategy for grouping of nanomaterials based on key physio-chemical descriptors as a basis for safer-by-design NMs. *Nano Today* **2014**, *9*, 266–270. [CrossRef]
28. Zhang, L.; Yu, J.; Park, B.H.; Kinzler, K.W.; Vogelstein, B. Role of BAX in the apoptotic response to anticancer agents. *Science* **2000**, *290*, 989–992. [CrossRef] [PubMed]
29. Panas, A.; Comouth, A.; Saathoff, H.; Leisner, T.; Al-Rawi, M.; Simon, M.; Seemann, G.; Dössel, O.; Mülhopt, S.; Paur, H.R.; et al. Silica nanoparticles are less toxic to human lung cells when deposited at the air-liquid interface compared to conventional submerged exposure. *Beilstein. J. Nanotechnol.* **2014**, *5*, 1590–1602. [CrossRef] [PubMed]
30. Hansjosten, I.; Rapp, J.; Reiner, L.; Vatter, R.; Fritsch-Decker, S.; Peravali, R.; Palosaari, T.; Joossens, E.; Gerloff, K.; Macko, P.; et al. Microscopy-based high-throughput assays enable multi-parametric analysis to assess adverse effects of nanomaterials in various cell lines. *Arch. Toxicol.* **2018**, *92*, 633–649. [CrossRef] [PubMed]
31. DeLoid, G.M.; Cohen, J.M.; Pyrgiotakis, G.; Demokritou, P. Preparation, characterization, and in vitro dosimetry of dispersed, engineered nanomaterials. *Nat. Protoc.* **2017**, *12*, 355–371. [CrossRef] [PubMed]
32. DeLoid, G.; Cohen, J.M.; Darrah, T.; Derk, R.; Rojanasakul, L.; Pyrgiotakis, G.; Wohlleben, W.; Demokritou, P. Estimating the effective density of engineered nanomaterials for in vitro dosimetry. *Nat. Commun.* **2014**, *5*, 3514. [CrossRef] [PubMed]

33. DeLoid, G.M.; Cohen, J.M.; Pyrgiotakis, G.; Pirela, S.V.; Pal, A.; Liu, J.; Srebric, J.; Demokritou, P. Advanced computational modeling for in vitro nanomaterial dosimetry. *Part Fibre Toxicol.* **2015**, *12*, 32. [CrossRef] [PubMed]
34. Donauer, J.; Schreck, I.; Liebel, U.; Weiss, C. Role and interaction of p53, BAX and the stress-activated protein kinases p38 and JNK in benzo(a)pyrene-diolepoxide induced apoptosis in human colon carcinoma cells. *Arch. Toxicol.* **2012**, *86*, 329–337. [CrossRef] [PubMed]
35. Dilger, M.; Orasche, J.; Zimmermann, R.; Paur, H.R.; Diabaté, S.; Weiss, C. Toxicity of wood smoke particles in human A549 lung epithelial cells: The role of PAHs, soot and zinc. *Arch. Toxicol.* **2016**, *90*, 3029–3044. [CrossRef]
36. Marquardt, C.; Fritsch-Decker, S.; Al-Rawi, M.; Diabaté, S.; Weiss, C. Autophagy induced by silica nanoparticles protects RAW264.7 macrophages from cell death. *Toxicology* **2017**, *379*, 40–47. [CrossRef]
37. Stöber, W.; Fink, A.; Bohn, E. Controlled growth of monodisperse silica spheres in the micron size range. *J. Colloid Interface Sci.* **1968**, *26*, 62–69. [CrossRef]
38. Mülhopt, S.; Diabaté, S.; Dilger, M.; Adelhelm, C.; Anderlohr, C.; Bergfeldt, T.; Gomez, D.L.T.; Jiang, Y.; Valsami-Jones, E.; Langevin, D.; et al. Characterization of Nanoparticle Batch-To-Batch Variability. *Nanomaterials* **2018**, *8*, 311. [CrossRef]
39. Al-Rawi, M.; Diabaté, S.; Weiss, C. Uptake and intracellular localization of submicron and nano-sized SiO_2 particles in HeLa cells. *Arch. Toxicol.* **2011**, *85*, 813–826. [CrossRef]
40. Kowoll, T.; Fritsch-Decker, S.; Diabaté, S.; Nienhaus, G.U.; Gerthsen, D.; Weiss, C. Assessment of in vitro particle dosimetry models at the single cell and particle level by scanning electron microscopy. *J. Nanobiotechnol.* **2018**, *16*, 100. [CrossRef]
41. Donaldson, K.; Brown, D.; Clouter, A.; Duffin, R.; MacNee, W.; Renwick, L.; Tran, L.; Stone, V. The pulmonary toxicology of ultrafine particles. *J. Aerosol Med.* **2002**, *15*, 213–220. [CrossRef] [PubMed]
42. Stoeger, T.; Schmid, O.; Takenaka, S.; Schulz, H. Inflammatory response to TiO_2 and carbonaceous particles scales best with BET surface area. *Environ. Health Perspect.* **2007**, *115*, A290–A291. [CrossRef] [PubMed]
43. Schmid, O.; Stoeger, T. Surface area is the biologically most effective dose metric for acute nanoparticle toxicity in the lung. *J. Aerosol Sci.* **2016**, *99*, 133–143. [CrossRef]
44. Morishige, T.; Yoshioka, Y.; Inakura, H.; Tanabe, A.; Narimatsu, S.; Yao, X.; Monobe, Y.; Imazawa, T.; Tsunoda, S.; Tsutsumi, Y.; et al. Suppression of nanosilica particle-induced inflammation by surface modification of the particles. *Arch. Toxicol.* **2012**, *86*, 1297–1307. [CrossRef] [PubMed]
45. Nowak, J.S.; Mehn, D.; Nativo, P.; Garcia, C.P.; Gioria, S.; Ojea-Jimenez, I.; Gilliland, D.; Rossi, F. Silica nanoparticle uptake induces survival mechanism in A549 cells by the activation of autophagy but not apoptosis. *Toxicol. Lett.* **2014**, *224*, 84–92. [CrossRef] [PubMed]
46. Slowing, I.I.; Wu, C.W.; Vivero-Escoto, J.L.; Lin, V.S. Mesoporous silica nanoparticles for reducing hemolytic activity towards mammalian red blood cells. *Small* **2009**, *5*, 57–62. [CrossRef] [PubMed]
47. Yu, T.; Malugin, A.; Ghandehari, H. Impact of silica nanoparticle design on cellular toxicity and hemolytic activity. *ACS Nano.* **2011**, *5*, 5717–5728. [CrossRef]
48. Andon, F.T.; Fadeel, B. Programmed cell death: Molecular mechanisms and implications for safety assessment of nanomaterials. *Acc. Chem. Res.* **2013**, *46*, 733–742. [CrossRef]
49. Galluzzi, L.; Bravo-San Pedro, J.M.; Vitale, I.; Aaronson, S.A.; Abrams, J.M.; Adam, D.; Alnemri, E.S.; Altucci, L.; Andrews, D.; Annicchiarico-Petruzzelli, M.; et al. Essential versus accessory aspects of cell death: Recommendations of the NCCD 2015. *Cell Death Differ.* **2015**, *22*, 58–73. [CrossRef]
50. Ye, Y.; Liu, J.; Xu, J.; Sun, L.; Chen, M.; Lan, M. Nano-SiO_2 induces apoptosis via activation of p53 and Bax mediated by oxidative stress in human hepatic cell line. *Toxicol. In Vitro* **2010**, *24*, 751–758. [CrossRef]
51. Ahmad, J.; Ahamed, M.; Akhtar, M.J.; Alrokayan, S.A.; Siddiqui, M.A.; Musarrat, J.; Al-Khedhairy, A.A. Apoptosis induction by silica nanoparticles mediated through reactive oxygen species in human liver cell line HepG2. *Toxicol. Appl. Pharmacol.* **2012**, *259*, 160–168. [CrossRef] [PubMed]
52. Galluzzi, L.; Joza, N.; Tasdemir, E.; Maiuri, M.C.; Hengartner, M.; Abrams, J.M.; Tavernarakis, N.; Penninger, J.; Madeo, F.; Kroemer, G. No death without life: Vital functions of apoptotic effectors. *Cell Death Differ.* **2008**, *15*, 1113–1123. [CrossRef] [PubMed]
53. Gerloff, K.; Fenoglio, I.; Carella, E.; Kolling, J.; Albrecht, C.; Boots, A.W.; Förster, I.; Schins, R.P. Distinctive toxicity of TiO_2 rutile/anatase mixed phase nanoparticles on Caco-2 cells. *Chem. Res. Toxicol.* **2012**, *25*, 646–655. [CrossRef] [PubMed]

54. Wiemann, M.; Vennemann, A.; Sauer, U.G.; Wiench, K.; Ma-Hock, L.; Landsiedel, R. An in vitro alveolar macrophage assay for predicting the short-term inhalation toxicity of nanomaterials. *J. Nanobiotechnol.* **2016**, *14*, 16. [CrossRef] [PubMed]
55. Dorier, M.; Beal, D.; Marie-Desvergne, C.; Dubosson, M.; Barreau, F.; Houdeau, E.; Herlin-Boime, N.; Carriere, M. Continuous in vitro exposure of intestinal epithelial cells to E171 food additive causes oxidative stress, inducing oxidation of DNA bases but no endoplasmic reticulum stress. *Nanotoxicology* **2017**, *11*, 751–761. [CrossRef]
56. Dorier, M.; Tisseyre, C.; Dussert, F.; Béal, D.; Arnal, M.E.; Douki, T.; Valdiglesias, V.; Laffon, B.; Fraga, S.; Brandao, F.; et al. Toxicological impact of acute exposure to E171 food additive and TiO_2 nanoparticles on a co-culture of Caco-2 and HT29-MTX intestinal cells. *Mutat. Res. Gen. Tox. Environ.* **2018**. [CrossRef]
57. Dorier, M.; Béal, D.; Tisseyre, C.; Marie-Desvergne, C.; Dubosson, M.; Barreau, F.; Houdeau, E.; Herlin-Boime, N.; Rabilloud, T.; Carriere, M. The food additive E171 and titanium dioxide nanoparticles indirectly alter the homeostasis of humanintestinal epithelial cells in vitro. *Environ. Sci. Nano* **2019**, *6*, 1549–1561. [CrossRef]
58. Zhou, M.; Huang, H.; Wang, D.; Lu, H.; Chen, J.; Chai, Z.; Yao, S.Q.; Hu, Y. Light-triggered PEGylation/dePEGylation of the nanocarriers for enhanced tumor penetration. *Nano Lett.* **2019**, *19*, 3671–3675. [CrossRef]

Article

Improving Quality in Nanoparticle-Induced Cytotoxicity Testing by a Tiered Inter-Laboratory Comparison Study

Inge Nelissen [1,*], Andrea Haase [2], Sergio Anguissola [3,4], Louise Rocks [3,5], An Jacobs [1], Hanny Willems [1], Christian Riebeling [2], Andreas Luch [2], Jean-Pascal Piret [6], Olivier Toussaint [6,†], Bénédicte Trouiller [7], Ghislaine Lacroix [7], Arno C. Gutleb [8], Servane Contal [8], Silvia Diabaté [9], Carsten Weiss [9], Tamara Lozano-Fernández [10,11], África González-Fernández [10,12], Maria Dusinska [13], Anna Huk [13,14], Vicki Stone [15], Nilesh Kanase [15], Marek Nocuń [16,17], Maciej Stępnik [16], Stefania Meschini [18], Maria Grazia Ammendolia [18], Nastassja Lewinski [19,20], Michael Riediker [19,21,22], Marco Venturini [23], Federico Benetti [23], Jan Topinka [24], Tana Brzicova [24,25], Silvia Milani [26], Joachim Rädler [26], Anna Salvati [3,27] and Kenneth A. Dawson [3]

1 Health Department, Flemish Institute for Technological Research (VITO), Boeretang 200, 2400 Mol, Belgium; an.jacobs@vito.be (A.J.); hanny.willems@vito.be (H.W.)
2 Department of Chemicals and Product Safety, German Federal Institute for Risk Assessment (BfR), Max-Dohrn-Strasse 8-10, 10589 Berlin, Germany; andrea.haase@bfr.bund.de (A.H.); christian.riebeling@bfr.bund.de (C.R.); andreas.luch@bfr.bund.de (A.L.)
3 Centre for BioNano Interactions, University College Dublin (UCD), Belfield, Dublin 4, Ireland; sergioanguissola@gmail.com (S.A.); Louise.rocks@sfi.ie (L.R.); a.salvati@rug.nl (A.S.); kenneth.a.dawson@cbni.ucd.ie (K.A.D.)
4 Charles River Laboratories, Carrowntreila, Ballina, Co. Mayo, Ireland
5 Science Foundation Ireland, Three Park Place, Hatch Street Upper, Dublin 2, Ireland
6 Research Unit in Cellular Biology (URBC), Namur Nanosafety Center (NNC), Namur Research Institute for Life Sciences (NARILIS), University of Namur (UNamur), rue de Bruxelles 61, 5000 Namur, Belgium; jean-pascal.piret@unamur.be
7 Experimental Toxicology Unit, Institut National de l'Environnement Industriel et des Risques (INERIS), Parc Alata, BP2, 60550 Verneuil-en-Halatte, France; benedicte.trouiller@ineris.fr (B.T.); ghislaine.lacroix@ineris.fr (G.L.)
8 Environmental Research and Innovation (ERIN) Department, Luxembourg Institute of Science and Technology (LIST), 41, rue du Brill, L-4422 Belvaux, Luxembourg; arno.gutleb@list.lu (A.C.G.); servane.contal@list.lu (S.C.)
9 Institute of Toxicology and Genetics, Karlsruhe Institute of Technology (KIT), Hermann-von-Helmholtz-Platz 1, 76344 Eggenstein-Leopoldshafen, Germany; silvia.diabate@kit.edu (S.D.); carsten.weiss@kit.edu (C.W.)
10 Biomedical Research Center (CINBIO), University of Vigo, Campus Lagoas Marcosende, 36310 Vigo, Spain; tlozano@nanoimmunotech.es (T.L.-F.); africa@uvigo.es (Ã.G.-F.)
11 Nanoimmunotech SL, Edificio CITEXVI Fonte das Abelleiras s/n, Campus Universitario de Vigo, 36310 Vigo, Pontevedra, Spain
12 Instituto de Investigación Sanitaria Galicia Sur (IISGS), Hospital Álvaro Cunqueiro, Estrada Clara Campoamor 341, Babio – Beade, 36312 Vigo, Spain
13 Health Effects Laboratory, Department of Environmental Chemistry, Norwegian Institute for Air Research (NILU), Instituttveien 18, 2007 Kjeller, Norway; maria.dusinska@nilu.no (M.D.); annahuk8@gmail.com (A.H.)
14 Gentian Diagnostics AS, Bjørnåsveien 5, 1596 Moss, Norway
15 School of Life Sciences, Heriot-Watt University (HWU), Riccarton Campus, Edinburgh EH14 4AS, UK; v.stone@hw.ac.uk (V.S.); n.kanase@hw.ac.uk (N.K.)
16 Department of Toxicology and Carcinogenesis, Nofer Institute of Occupational Medicine (NIOM), 91-348 Łódź, Poland; mareknocun@gmail.com (M.N.); mstep@imp.lodz.pl (M.S.)
17 SEQme s.r.o., Dlouha 176, 26301 Dobris, Czech Republic

[18] National Center for Drug Research and Evaluation and National Center of Innovative Technologies for Public Health, Istituto Superiore di Sanità (ISS), Viale Regina Elena, 299 Rome, Italy; stefania.meschini@iss.it (S.M.); maria.ammendolia@iss.it (M.G.A.)
[19] Institute for Work and Health (IST), University of Lausanne and University of Geneva, Route de la Corniche 2, 1066 Epalinges-Lausanne, Switzerland; nalewinski@vcu.edu (N.L.); michael.riediker@alumni.ethz.ch (M.R.)
[20] Department of Chemical and Life Science Engineering, Virginia Commonwealth University, Richmond, VA 23284, USA
[21] Swiss Centre for Occupational and Environmental Health (SCOEH), Binzhofstrasse 87, 8404 Winterthur, Switzerland
[22] School of Materials Science & Engineering, Nanyang Technological University, Block N4.1, Nanyang Avenue, Singapore 639798, Singapore
[23] ECAMRICERT SRL, European Center for the Sustainable Impact of Nanotechnology (ECSIN), Corso Stati Uniti 4, 35127 Padova, Italy; m.venturini@ecamricert.com (M.V.); f.benetti@ecamricert.com (F.B.)
[24] Institute of Experimental Medicine (IEM), Czech Academy of Sciences, Videnska 1083, 14220 Prague 4, Czech Republic; jtopinka@iem.cas.cz (J.T.); tana.brzicova@iem.cas.cz (T.B.)
[25] Faculty of Safety Engineering, VSB-Technical University of Ostrava, Lumirova 13, 70030 Ostrava-Vyskovice, Czech Republic
[26] Faculty of Physics and Center for NanoScience, Ludwig-Maximilians-Universität, Geshwister-Scholl-Platz 1, 80539 Munich, Germany; s.milani77@googlemail.com (S.M.); raedler@lmu.de (J.R.)
[27] Groningen Research Institute of Pharmacy, University of Groningen, A. Deusinglaan 1, 9713AV Groningen, The Netherlands
* Correspondence: inge.nelissen@vito.be; Tel.: +32-14-335107
† This work is dedicated to Prof. Toussaint who unfortunately deceased in 2016.

Received: 25 June 2020; Accepted: 17 July 2020; Published: 22 July 2020

Abstract: The quality and relevance of nanosafety studies constitute major challenges to ensure their key role as a supporting tool in sustainable innovation, and subsequent competitive economic advantage. However, the number of apparently contradictory and inconclusive research results has increased in the past few years, indicating the need to introduce harmonized protocols and good practices in the nanosafety research community. Therefore, we aimed to evaluate if best-practice training and inter-laboratory comparison (ILC) of performance of the 3-(4,5-dimethylthiazol-2-yl)-5-(3-carboxymethoxyphenyl)-2-(4-sulfophenyl)-2H-tetrazolium (MTS) assay for the cytotoxicity assessment of nanomaterials among 15 European laboratories can improve quality in nanosafety testing. We used two well-described model nanoparticles, 40-nm carboxylated polystyrene (PS-COOH) and 50-nm amino-modified polystyrene (PS-NH2). We followed a tiered approach using well-developed standard operating procedures (SOPs) and sharing the same cells, serum and nanoparticles. We started with determination of the cell growth rate (tier 1), followed by a method transfer phase, in which all laboratories performed the first ILC on the MTS assay (tier 2). Based on the outcome of tier 2 and a survey of laboratory practices, specific training was organized, and the MTS assay SOP was refined. This led to largely improved intra- and inter-laboratory reproducibility in tier 3. In addition, we confirmed that PS-COOH and PS-NH2 are suitable negative and positive control nanoparticles, respectively, to evaluate impact of nanomaterials on cell viability using the MTS assay. Overall, we have demonstrated that the tiered process followed here, with the use of SOPs and representative control nanomaterials, is necessary and makes it possible to achieve good inter-laboratory reproducibility, and therefore high-quality nanotoxicological data.

Keywords: nanosafety; cytotoxicity; inter-laboratory comparison; best practice; training

1. Introduction

Within the last 20 years, there has been a tremendous increase in numbers of publications on nanomaterial (NM) toxicity, many of which report inconclusive and controversial results, often apparently conflicting. This has generated a wide debate on the quality and relevance of published papers in nanosafety, including—for instance—a discussion opened up by the *Nature Nanotechnology* journal [1,2], which was followed by several commentaries and other examples [3,4]. A similar discussion about the reliability and reproducibility of experimental data has also been raised for science in general [5–7], as it was demonstrated for pre-clinical studies on cancer [8]: using in-house data trying to validate the published results, the authors found that at most only 25% were in line with published data. Efforts to adhere to the biological models as used in the original publications did not improve these results. Conversely, reproducible results were also transferable between models [8]. Largely, discrepancies arise from honest mistakes and flawed statistics, but a recent focus on bad practices and fraud in science has also uncovered instances of the latter [9,10]. Most of the published research has been performed in research laboratories, e.g., at universities, which usually do not adhere to good laboratory practice (GLP) or similar standards. GLP was developed specifically out of an experience of data manipulation and fraud in toxicological contract research [11–13]. In regulatory toxicology, adherence to GLP, Organization for Economic Cooperation and Development (OECD) guidelines and International Organization for Standardization (ISO) standards is therefore paramount. Applying similar standards to the existing literature on nanotoxicology by checking publications against a defined set of criteria, such as physico-chemical material characterization or detailed descriptions of the assays applied, including solid statistical data evaluation, resulted in approximately 68% [14] to 90% [15] of the studies being rejected.

Even if it sounds trivial, it has to be stressed that in nanotoxicology all steps from synthesis route to nanoparticle (NP) sample preparation for testing and every step in-between will have an effect on the outcome of the experiments. NMs may become altered throughout all these processes. For biological testing, the dispersion of NMs into media has two crucial aspects, the type of dispersion [16] and the composition of the media [17]. The energy used for the dispersion can, for instance, passivate the surface, influence the agglomeration state and cause dissolution of molecules into ions [18]. Liquid media and biological fluids can influence the same parameters and, importantly, the presence of proteins and other biomolecules leads to the formation of a biocorona on the surface of the NMs [19]. The impact of the corona on NP-cell interactions is demonstrated by the differential cytotoxicity of NPs in the absence or presence of serum, as a general paradigm for all NMs, such as for instance silica [20,21], positively charged NPs [22], carbon nanotubes [23] and graphene oxides [24]. While under realistic exposure conditions in biological fluids the NMs are passivated by the presence of this layer of biomolecules from the surrounding environment, in artificially simplified laboratory conditions, such as serum-free medium, the bare surfaces of NMs can adhere so strongly to the cell surface that they generate damage and other biological processes. Importantly, even the amount and identity of proteins present affects NP outcome on cells [25–27], opening up new challenges for determining realistic exposure scenarios. Other unique features of NMs affecting the outcomes of toxicity testing in comparison to standard chemicals include the interference of the NM itself with the testing method: NMs can adsorb and scatter light, interfering with tests based on absorbance, luminescence and fluorescence detection. Furthermore, NMs can also adsorb the reagents used for the tests on their surface, thus causing artefacts that could be misinterpreted for signals [28]. These are just some examples of the many unique features of materials at the nanoscale, which have caused the need for the development of new methods and procedures, as well as specific laboratory practices, in order to be able to generate robust and reproducible data in nanosafety.

These issues are the main reasons why most studies are incomparable [4]. To achieve improved comparability and reproducibility, protocols have to be harmonized and standardized. Moreover, there is a need for the development of alternative testing strategies, as the vast possibilities of engineering NMs would result in a high number of animal studies if current regulatory protocols were to be followed. Animal studies are not only costly, time consuming and ethically fraught; the legislation of cosmetics in the European Union already prohibits this type of testing. Alternative in vitro testing might be a useful tool to prioritize animal testing to NMs of concern. In addition, these methods might be used during R&D to eliminate substances with hazardous properties. Standardization of in vitro procedures requires examining all involved materials and specifying every single step in a protocol. For instance, cell lines have to be identical [29] and free of mycoplasma [30], and even the way how cells are seeded into a multi-well plate has consequences on the test result [31].

In this study, an inter-laboratory comparison (ILC) study among 15 European research laboratories was organized to test how easy it was to generate reproducible data on NP-induced toxicity using standard operating procedures (SOPs), and define processes to enhance the proficiency of nanosafety research laboratories in achieving reliable NM toxicity data. We have employed the in vitro 3-(4,5-dimethylthiazol-2-yl)-5-(3-carboxymethoxyphenyl)-2-(4-sulfophenyl)-2H-tetrazolium (MTS) cytotoxicity assay of the tetrazolium salt reduction-type [32] as a benchmark, which has been addressed before by other ILC consortia consisting of 8 [33], 5 [34] or 6 [35] independent laboratories. Briefly, SOPs were generated by laboratories with previous experience on this assay, as well as in working according to the principles of GLP. The SOP for the MTS assay, including the 96-well plate lay-out and performance criteria, was in agreement with Rösslein et al. [36] and the recently published ISO 19007:2018 standard [37]. Cells, serum and NPs, all from the same batches, were shared among the participating laboratories across Europe, which were enrolled based on the outcome of a first tier on cell culturing proficiency. The second tier was a first MTS inter-laboratory comparison study. Its results, followed by a questionnaire sent to all participants to collect more information on how the procedure was followed, clearly highlighted the need for further optimization of the developed SOP, but also the need of a more precise training in executing this kind of standardized testing. After training of the participants, the third tier consisted of a new round of the MTS assay using the revised SOP. The final results showed a strong reduction in the variability within and across laboratories. Furthermore, our data endorsed the potential of amine-modified polystyrene nanoparticles (PS-NH2) and carboxyl-modified polystyrene nanoparticles (PS-COOH) as positive and negative control nanomaterials, respectively, for cytotoxicity assessment using the MTS assay.

2. Materials and Methods

2.1. Recruitment of Laboratories

Laboratories involved in nanosafety research across Europe that were (associated) partners of the QualityNano Research Infrastructure consortium were invited to the inter-laboratory comparison study. Before the start of the study, an online questionnaire was sent out to the candidate laboratories to inquire about and evaluate them against criteria mentioned in ISO/IEC 17043:2010 'Conformity assessment—general requirements for proficiency testing' [38], including: experience in biological assessments of NMs or in performing tests similar to the proposed method and/or cell model; the availability of the technical requirements for accommodation, environmental conditions and endpoint measurements; high quality standards for biological testing implemented, such as good cell culture practice or GLP; and trained personnel. In total, 15 laboratories from academia (33.3%), research organizations (60.0%) and industry (6.7%) joined the study. The majority of them were not familiar with the requested high-quality standards that have been developed primarily for regulatory testing or method validation. Moreover, the test performers had varying qualifications (lab technician, PhD student or post-doctoral scientist) and a varied numbers of years of experience in biological testing (from a few months to over 20 years). In most cases, they were not trained in the proposed SOPs.

2.2. Choice of Cytotoxicity Test and Materials Used

As mentioned above, we selected the in vitro MTS cytotoxicity assay as a benchmark assay to evaluate and improve laboratories' proficiency. The CellTiter 96® Aqueous One assay (Promega,

Leiden, The Netherlands) used in this study has been identified as superior to other cell viability assays for NP assessment [39]. The assay has originally been developed by Tim Mosmann [32] for the measurement of cell viability using 3-(4,5-dimethylthiazol-2-yl)-2,5-diphenyltetrazolium bromide (MTT). Soluble tetrazolium salts, including 3-(4,5-dimethylthiazol-2-yl)-5-(3-carboxymethoxyphenyl)-2-(4-sulfophenyl)-2H-tetrazolium (MTS), have later been developed, of which the tetrazolium ring is reduced by cellular nicotinamide adenine dinucleotide (phosphate)-dependent oxidoreductase enzymes in the presence of an intermediate electron acceptor (phenazine ethosulfate), to form a formazan derivative that is quantified in a spectrophotometer. The product therefore reflects the metabolic activity and by extension viability of cells, and, hence, can be used to determine a toxic dose of a substance. Information on the cytotoxicity of a substance in alternative methods using cell culture is a crucial first screening step to any more detailed investigation, or as part of a safe-by-design approach. Test methods to assess the in vitro cytotoxicity of medical devices have been described in ISO 10993-5:2009 [40]. A standard specifically dedicated to nanomaterials using the MTS assay as an in vitro cytotoxicity assay has more recently been published [37].

The human A549 alveolar epithelial cell line was chosen as a cell type representing the respiratory route of exposure and a major site of deposition of small nanoparticles [41,42]. Moreover, this cell model is widely used in nanosafety laboratories and easy to maintain, and therefore suitable for standardization among different laboratories.

Alternative methods for toxicity testing require substances with known potential, either positive or negative control substances, which are crucial as quality controls in SOPs. So far, several NM controls have been suggested for in vitro nanoparticle toxicity studies, including tungsten carbide-cobalt [43] as positive, and barium sulfate [44] and carboxylated nanodiamonds [45] as negative controls. As is true for chemicals, any NM that would be used as a control for a specific assay needs to be thoroughly characterized with respect to its physico-chemical properties, and its performance in this assay also needs to be well described. However, for NMs special care has to be taken, because of possible batch-to-batch variation, contamination and long-term stability [46]. In our study, we have selected 50-nm amine-modified polystyrene NPs (PS-NH2) and 40-nm carboxyl-modified polystyrene NPs (PS-COOH), which are known to form stable dispersions when diluted in cell culture medium [47–49]. Furthermore, their impact on cells has been characterized in detail [22,26,48,50–52]. Thus, they constituted ideal candidates as starting materials for this ILC study.

2.3. Standardization Procedures and SOP Development

Standard operating procedures for cell culturing and cell growth rate determination (tier 1), and cytotoxicity assessment using the MTS assay (tier 2 and 3) were adopted from existing protocols, and adapted to implement the spirit of Good Cell Culture Practice [53] and GLP [13], and to be in line with the ISO 19007:2018(E) standard [37]. Forms for detailed registration of performance of the protocol steps were prepared and filled in by the partner laboratories, to make it possible to formulate corrective actions in case of deviations in a laboratory's results. For data analysis and reporting, spreadsheet (Microsoft Excel) templates and web forms were prepared and distributed to the laboratories. These enabled automated calculations and immediate evaluation of compliance with the test acceptance criteria as formulated in the SOPs. After a first ILC on the MTS assay (tier 2) with the developed SOPs, further changes were made to the MTS assay SOP, and forms based on the feedback from the participating laboratories through an online questionnaire. The optimized SOPs and the spreadsheet templates are available in Supplementary Materials.

To enhance standardization in cell growth and test performance, all laboratories used the most critical materials, such as fetal bovine serum and test NPs from a centrally held stock prepared at one location. Fresh aliquots were shipped to the laboratories prior to the start of the studies. In addition, identical frozen cell stocks of the human A549 alveolar epithelial cell line were obtained by the participating laboratories from a central laboratory, which purchased a parent cell line (ATCC, CCL-185, passage number 82) and subcultured the cells up to a master cell bank (passage number

88). A harmonized SOP for thawing, freezing and subculturing of the A549 cell line, and testing for mycoplasma contamination according to an in-house protocol as an essential quality control before freezing the cells, was developed and used by the study partners to generate their own working cell bank. Other critical reagents, such as cell culturing reagents, MTS reagent and staurosporine were used from the same supplier by the laboratories.

2.4. Nanoparticles and Chemical Control

PS-NH2 (50 nm) and fluorescently labelled PS-COOH (40 nm) were purchased from Bangs Laboratories Inc. (Fishers, IN, USA; catalogue number PA02N) and Molecular Probes (ThermoFisher Scientific, Bio-Sciences, Dublin, Ireland; catalogue number F8795), respectively. Aliquots of diluted NPs (10 mg/mL) in milliQ water (resistivity of 18.2 mΩ.cm at 25 °C) were prepared centrally, distributed among all participants, and stored at 4 °C. Dispersions of the NPs of 100 µg/mL in complete cell culture medium (CCM) containing minimal essential medium with GlutaMAX™ (Gibco®, Life Technologies, Paisley, UK), supplemented with 10% (*v*/*v*) non-heat inactivated fetal bovine serum (FBS; Gibco®, Life Technologies, Paisley, UK), 100 U/mL penicillin and 100 µg/mL streptomycin (Gibco®, Invitrogen, Paisley, UK) were prepared by each laboratory. This was done by pipetting 20 µL of the NP stock suspensions in 1980 µL medium and mixing on a vortex for 30 s.

Staurosporine (Proteinkinase, Biaffin GmbH & Co., KG, Kassel, Germany; catalogue number PKI-STSP-001) was used as a positive chemical control to serve as an internal control of the biological cell response and MTS assay performance. As this compound was observed to become instable and lose its activity during transport and storage when prepared as solution, all laboratories were asked to purchase their own lot of lyophilized powder from the same company and with the same batch number. A stock solution of 1 mM staurosporine was prepared by dissolving the powder in 214 µL dimethyl sulfoxide (DMSO) and mixing on a vortex, followed by immediate further dilution to 2140 µL DMSO. Aliquots of 50 µL stock solution were stored at −80 °C without loss of activity for at least 6 months, and a fresh tube was thawed for each experiment. Solutions of 1000 nM staurosporine were prepared in complete CCM containing 1% milliQ water to ensure identical vehicle to the NPs. Details on the test item preparations are described in the SOP for the MTS cytotoxicity assay (Supplementary Materials).

2.5. Cell Culture and Exposure

Human A549 alveolar epithelial cells were maintained in complete CCM without antibiotics. The cells were subcultured every 3–4 days when the cell monolayer reached 70–80% confluence, with medium renewal after 2 days. Cells were used for testing up to passage number 20, to ensure equally low passage numbers among experiments and laboratories. Cells were used at >90% viability. Details are in the SOP for A549 cell culturing in Supplementary Materials.

Cell growth curves were obtained as detailed in the SOP for assessment of A549 cell growth rate and viability (Supplementary Materials) by determining the cell number and viability by trypan blue exclusion at 24 h, 48 h and 72 h after seeding.

Assessment of NP-induced cytotoxicity by the MTS assay was performed in 96-well plates, in which 200 µL of cell suspension were seeded per well, followed by 24 h incubation at 37 °C, 5% CO_2 to allow for cell adhesion prior to exposure to the test items. The dosing plate layout, shown in Figure S1, contained three replicate series of 6 concentrations of either the PS-NH2 or PS-COOH NPs (0; 1; 10; 25; 50 and 100 µg/mL) and 6 concentrations of staurosporine (0; 62.5; 125; 250; 500; and 1000 nM) as a positive chemical control, in addition to six replicate wells for each of 4 assay controls, according to Rösslein et al. [36]. Triplicate dilution series of the NPs and staurosporine were each started from a separate preparation of the highest test concentration, in order to estimate within-laboratory variability in the preparation. Dose series of the test items were obtained by serial dilutions in complete CCM supplemented with antibiotics, and containing 1% milliQ water to ensure identical vehicle in each well. Assay controls consisted of untreated cells in complete CCM with 1% milliQ water, blank wells (no cells) containing complete CCM with 1% milliQ water, and blank wells (no cells) containing the highest test concentrations of the NPs and staurosporine to control for potential interference with the assay read-out measurement. Cells were exposed by the removal of the medium from each well and transfer of 100 µL of the chemical and NP doses from the dosing plate, and, again, incubated for 24 h. All details and plate layouts can be found in the SOP for the MTS cytotoxicity assay (Supplementary Materials).

2.6. MTS Assay

At the end of the incubation of cells with the NPs and staurosporine doses, the medium was removed from each well and replaced with 150 µL of diluted CellTiter 96® AQueous One Solution Reagent containing MTS and the electron coupling reagent phenazine ethosulfate (Promega, Leiden, The Netherlands). Cell plates were incubated at 37 °C and 5% CO_2 for 1 h, to allow for bioreduction of the tetrazolium compound to a colored, soluble formazan product. The absorbance of the product at 490 nm, which is directly proportional to the number of living cells in culture was recorded using a spectrophotometer. Measurements were performed in the presence of the cells, and—after SOP optimization—after transfer of 100 µL of the colored reagent from the cell plate into a new 96-well plate, to ensure absorbance read-out in the linear dynamic range relevant to the Lambert-Beer law.

2.7. Statistical Data Analysis and Proficiency Testing

At least 3 independent runs executed on different days using cell cultures of different passage numbers were performed for cell growth rate determination (tier 1) and the MTS assay (tier 2 and 3). An initial analysis of the raw data was done by the individual laboratories using a spreadsheet calculation template enabling immediate evaluation of compliance with the test acceptance criteria mentioned in the SOPs, as well as automated data analysis and plotting (Supplementary Materials). Only data sets from fully acceptable tests were considered valid to be included in the final statistical analysis.

In the determination of cell growth rate, the total cell counts and percentages of cell viability were calculated, based on the live and dead cell numbers. A curve presenting the live cell counts at different growth times (24, 48 and 72 h) was used for exponential fitting (Figure S2). The prefactor of the exponential power from the resulting equation indicated the relative growth rate (doublings per hour), and was used to calculate the cell doubling time (hours).

In a run of the MTS assay, the percentage cell survival was calculated as the fraction of cells that remained viable after treatment, by subtracting the average background absorbance of medium blank wells (Figure S1, column 7) from each raw absorbance value, and normalizing the resulting values to the average absorbance of untreated cells. The triplicate values from a single dose were used to calculate a mean value and standard deviation (SD).

The statistical analysis of the MTS assay data from all laboratories was automated using in-house programming in R software [54]. Fitted sigmoidal curves with the upper limit fixed to 100 and lower limit to 0 were generated per run based on the dose-response data, and used to calculate the effective concentrations causing 30% (EC30, for staurosporine) or 50% (EC50, for PS-NH2) inhibition of cell viability. For curve fitting the R package drc was used [55] and the four-parameter logistic model with parameters b, c, d, e:

$$f(x, (b, c, d, e)) = c + (d - c)/(1 + \exp[b(\log(x) - \log(e))]) \quad (1)$$

Parameter e corresponds to the EC30 or EC50, whereas parameter b denotes the relative slope around e. The logistic function is symmetric around e. Examples of resulting dose-response curves per run of two different laboratories are shown in Figure S3.

For proficiency testing by inter-laboratory comparisons, statistical methods according to ISO 13258:2005(E) were followed in this study. More specifically, a robust statistical approach described in algorithm A [56] was applied to calculate robust values of the mean and SD of cell doubling times, % cell survival per dose, and EC30 (or EC50) from at least 3 independent runs reported by the individual participants in a round of the proficiency testing scheme. Next, the robust overall mean and SD from the data of all laboratories were calculated, starting from the robust mean data per lab using the algorithm A. Additionally, for the MTS data, based on the robust within-laboratory SD for each dose or EC30 (or EC50), a robust overall SD* was calculated using algorithm S [56], which yields a robust pooled value of the SD values, to which it is applied. Intra- and inter-laboratory biases were interpreted on the basis of overall mean and SD values derived from all laboratories. We concluded that a laboratory was proficient if the robust within-laboratory mean value did not exceed the overall mean with more than 2-fold the overall SD. Similarly, if the SD values of one laboratory were within the range of twice the overall SD*, then the laboratory bias at the individual dose or EC30 (or EC50) level was considered acceptable.

Finally, a coefficient of variation (CV) was calculated to evaluate the reproducibility of assay performance as: (robust SD/robust mean) × 100%. Variability of biological test results within or between laboratories was considered acceptable if CV < 30%.

3. Results

3.1. Determination of Cell Growth Rate (Tier 1)

In a first tier of the study, an SOP on A549 cell culturing and an SOP on determination of cell growth rate and viability (Supplementary Materials) were used by the 15 participating laboratories, to assess variability in the cell growth characteristics. Cell cultures for independent runs were started from independent cell vials of the working cell bank and tested for the absence of mycoplasma infection using in-house procedures. Relative cell viability was tested by trypan blue exclusion and was observed to fulfil the acceptance criterion (>90%) mentioned in the cell culturing SOP in the different laboratories. The mean cell doubling time derived from individual cell growth curves of all laboratories was 24.9 ± 2.4 h, and showed good agreement within and between the participating laboratories. Each laboratory produced at least two independent measurements, resulting in a mean doubling time within the boundaries of the overall mean ± 2SD, which indicated that they were all proficient in cell culturing (Figure 1). The largest variability was observed within laboratory 15, which obtained a mean cell doubling time of 28.4 ± 3.2 h.

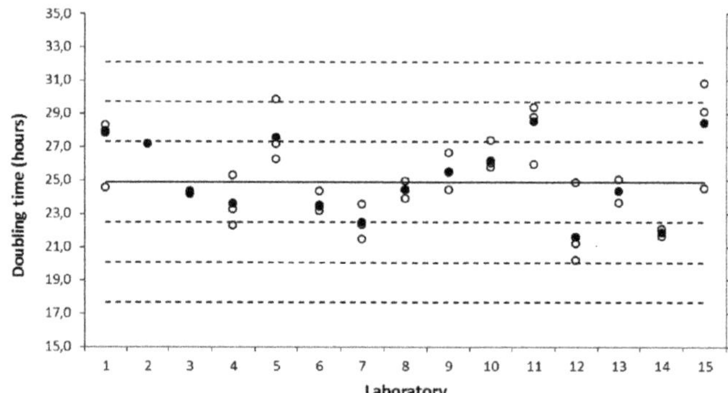

Figure 1. Inter-laboratory evaluation of growth rate of the A549 cell line (tier 1). Three (some 2) independent runs (indicated with open circles) were performed in each laboratory to determine the cell doubling time. The mean of each lab (filled circles), and overall mean (N = 15; black solid line) with 1-, 2- or 3-fold of the overall standard deviation (SD) (black dotted lines) are indicated.

3.2. Assessment of Laboratories' Inherent Proficiency in Performing the In Vitro MTS Assay (Tier 2)

After benchmarking and confirming the proficiency of the laboratories in their cell culturing performance, they were enrolled in a second tier of the inter-laboratory comparison study involving the transfer of the SOP on the MTS assay. Here, we aimed to evaluate the inherent proficiency of each participating laboratory, and identify critical phases in the SOP that introduce bias in testing, to allow further optimization of the SOP. For this study tier, the laboratories performed independent runs using cells of different passage number from the same working cell bank vial. Results on cytotoxicity assessment of the positive chemical control staurosporine showed that eight out of 15 laboratories reported on at least three independent and valid runs, based on assessment against the test acceptance criteria mentioned in the SOP. For the tests with PS-COOH and PS-NH2, however, only five and four out of 15 participating laboratories, respectively, were able to generate at least three valid runs (Table 1). The most frequent reason for this low rate of valid runs was a deviation of more than 15% in the absorbance values from triplicate cultures treated with zero dose of the test item (staurosporine or NPs; Figure S1, row B), as compared to untreated cultures (Figure S1, column 6). This can be attributed to the differences in seeded cell numbers or cell densities in the respective wells of the multi-well plate, which may have multiple causes, such as the poor resuspension of cells while seeding, inaccurate pipetting volumes, wrong pipetting technique, etc. Furthermore, about half of the laboratories observed interference of the PS-NH2 with the absorbance read-out for more than 15% compared to blank wells containing no NPs, which was set as a limit for acceptance of the test. More detailed data investigation revealed that this was mainly due to high variability (CV >30%) between replicate wells of the NP blank within these laboratories.

Table 1. Compliance of laboratories with the acceptance criteria of the 3-(4,5-dimethylthiazol-2-yl)-5-(3-carboxymethoxyphenyl)-2-(4-sulfophenyl)-2H-tetrazolium (MTS) cytotoxicity assay. The number of compliant laboratories compared to the total number of laboratories is indicated per test item.

Acceptance Criteria	Staurosporine	PS-COOH	PS-NH2
- Average absorbance values of NP blank deviating <15% from medium blank (no NP interference)	n/a [1]	14/15	8/15
- Blank replicate values, CV [2] <30%	15/15	15/15	15/15
- Blank-corrected absorbance values >0.1	15/15	15/15	14/15
- Blank-corrected absorbance values, CV <30%	12/15	14/15	6/15
- Average absorbance values at zero dose deviating <15% from non-treated cells	8/15	6/15	7/15
- % cell survival <70% for at least one concentration of positive chemical control (staurosporine)	13/15	14/15	13/15
≥ three valid and independent runs	**8/15**	**5/15**	**4/15**

[1] n/a, not applicable; [2] CV, coefficient of variation.

The dose-response data of staurosporine and NP-induced cytotoxicity in A549 cells were evaluated using robust statistical methods. Examples of dose-response data from single laboratories are shown in Figure S3. Dose-dependent decrease in percentage cell survival was observed for staurosporine and PS-NH2, whereas PS-COOH did not affect cell viability as expected. Intra- and inter-laboratory biases were calculated, while including either all data from both valid and non-valid runs of all laboratories (Figure S4), or only the data from the laboratories with at least three valid runs (Figure 2).

Overall variability in the dose-response data, represented by the SD of the mean % cell survival of all laboratories and the SD*, which represents a pooled value of the within-laboratories' SD, was observed to decrease when non-valid runs were discarded, indicating improved inter-laboratory reproducibility. At the same time, the number of laboratories with an intra-laboratory bias exceeding the overall mean plus 2-fold of the overall SD decreased (Figure 2 vs. Figure S4). These findings highlight the need to apply acceptance criteria as quality measures to the biological test performance to enhance reproducibility of results and hence standardization.

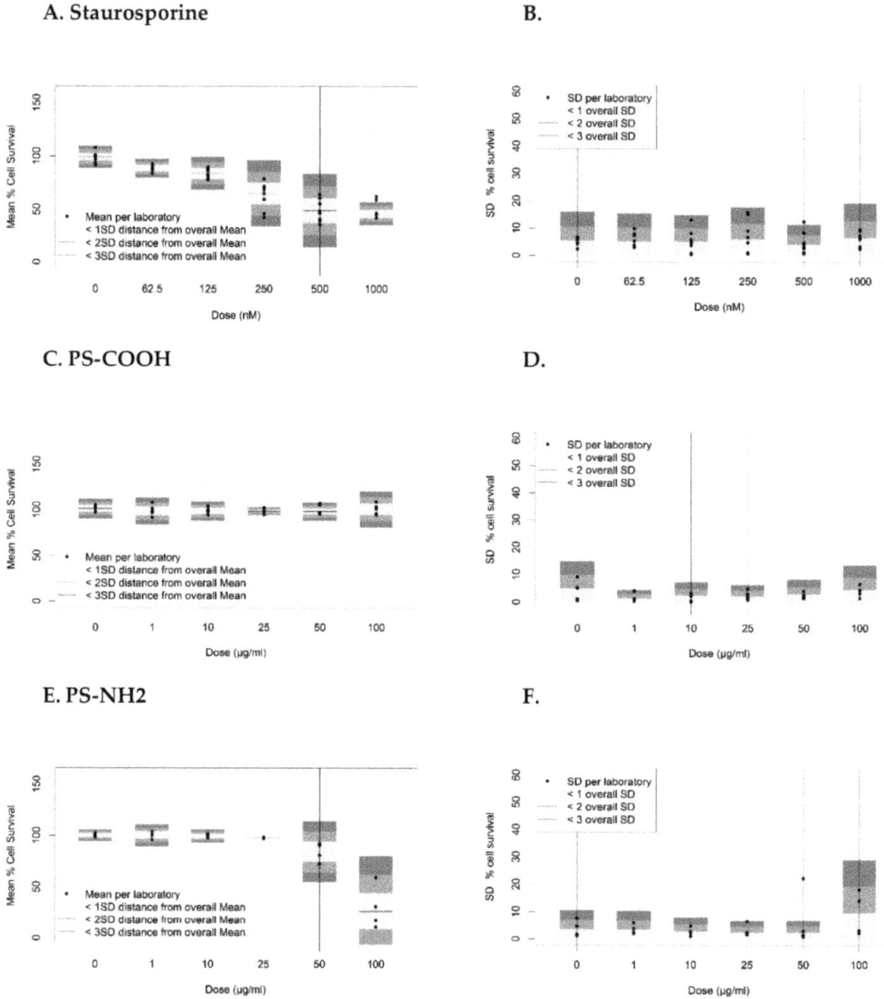

Figure 2. Intra- and inter-laboratory biases for percentage cell survival determined using the MTS assay based on valid runs (tier 2). Mean percentage cell survival per dose and per laboratory compared to the overall mean and SD (left panels), as well as SD of percentage cell survival per dose and per laboratory compared to the overall SD* (right panels) are shown for (**A**,**B**) staurosporine (N = 8), (**C**,**D**) carboxyl-modified polystyrene nanoparticles (PS-COOH) (N = 5) and (**E**,**F**) amine-modified polystyrene nanoparticles (PS-NH2) (N = 4). Horizontal bars (left panel) indicate overall mean values of percentage cell survival, while grey shaded areas indicate the distances from the overall mean corresponding to 1-, 2- or 3-fold the overall SD (left panels) or SD* (right panels).

To further assess the intra- and inter-laboratory bias of toxicity values obtained in the different laboratories, EC30 and EC50 values were derived from the fitted dose-response curves resulting from cell exposures to staurosporine and PS-NH2, respectively. Although an EC50 value can usually be derived more accurately than an EC30 value, approximately half of the participants were not able to observe 50% inhibition of cell viability at 1000 nM staurosporine. Therefore, the EC30 of staurosporine was reported and compared, since it corresponds to the threshold value for concluding on cytotoxicity according to ISO 10993-5:2009 [40]. When both non-valid and valid runs were taken into account, mean fitted EC30 of 329.7 ± 182.0 nM staurosporine and EC50 of 76.2 ± 20.4 µg/mL PS-NH2 were observed, with respective CVs of 55.2% for staurosporine and 26.8% for PS-NH2. These effective concentrations changed to an EC30 of 261.0 ± 121.1 nM staurosporine and EC50 of 76.2 ± 12.9 µg/mL PS-NH2 (Figure 3A,C) when only valid runs were considered, resulting in decreased CVs of 46.4% for staurosporine and 16.9% for PS-NH2, respectively. By excluding non-valid data the overall variability SD* of the EC50 values determined for PS-NH2 exposed cells was again decreased (10.6 for non-valid and valid data, vs. 6.9 for valid data; Figure 3D), indicating that intra-laboratory performances were also improved. In contrast, this improvement was not observed for EC30 determinations in tests with staurosporine (SD* of 78.7 vs. 84.6; Figure 3B), which, in general, showed lower reproducibility within and between laboratories. The latter is in agreement with the low stability of staurosporine in solution we observed during the study, and, therefore, other stable compounds, such as cadmium sulfate [34,37] are more suitable as a positive control.

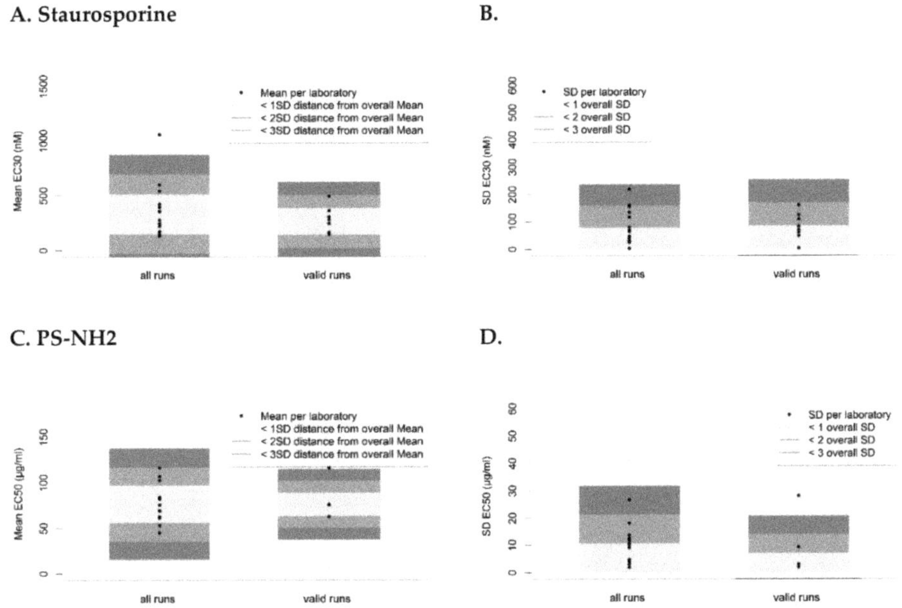

Figure 3. Intra- and inter-laboratory biases for effective concentration causing 30% inhibition of cell viability (EC30) and effective concentration causing 50% inhibition of cell viability (EC50) values determined using the MTS assay (tier 2). Mean values for EC30 and EC50 per laboratory compared to the overall mean and SD (left panels), as well as SD of EC30 and EC50 values per laboratory compared to the overall SD* (right panels) are shown for (**A**,**B**) staurosporine (N = 15 for all runs, N = 8 for valid runs) and (**C**,**D**) PS-NH2 (N = 15 for all runs, N = 4 for valid runs). Horizontal bars (left panel) indicate overall mean values of percentage cell survival, while grey shaded areas indicate the distances from the overall mean corresponding to 1-, 2- or 3-fold of the overall SD (left panels) or SD* (right panels).

Although variabilities were, in most cases, decreased when only valid results were included, these were still high (at the limit of what is acceptable) and warranted further investigation of possible causes and refinement of the benchmarking process. Thus, we examined in more detail if these first MTS experiments could highlight sensitive steps in the applied procedure, which introduced variability in the reported outcomes. This was done by collecting feedback on the interpretation of the SOP and performance of the MTS test by the participating laboratories using an online questionnaire. A list of critical steps that were reported in this survey is included in Table 2. These steps, including accurate pipetting volumes and seeding cell densities, were found to be quite similar to those already described in Rösslein et al. [36] and Elliott et al. [34].

Table 2. Critical steps in the standard operating procedure (SOP) for the MTS cytotoxicity assay. Following the analysis of the first data obtained for cells treated with staurosporine, PS-NH2 and PS-COOH, and the collection of responses of the participating laboratories to an online questionnaire on SOP interpretation and test performance, sensitive steps in the SOP that could generate and explain the observed variability of outcomes have been individuated.

Protocol Step	Critical Phase
All steps	- Verification of pipets and instruments - Use of single vs. multi-channel pipets - Pipetting technique - Adherence to timings stated in the SOP
Preparation and storage of staurosporine stock	- Dissolution of lyophilized product
Preparation of staurosporine working solution	- Low pipetting volume
Preparation of NP dilutions in cell culture medium	- Low pipetting volume - Dispersion protocol
Preparation of dosing plate	- Different pipetting volumes
Plating cells	- Use of antibiotics-free cell culture medium - Cell counting method - Homogeneous suspension of cells - Edge effects
Exposure to test item	- Removal of medium from cultures - Homogeneous suspension of test items - Application of test solutions onto cultures
MTS assay	- Removal of medium from cultures - Air bubbles - Precipitate of MTS reagent - Transfer of MTS reagent for read-out - Spectrophotometer specifications

Based on a thorough examination of the data and statistical analysis, and the collected feedback on critical phases in the test performance, we further optimized the SOP of the MTS cytotoxicity assay (available in the Supplementary Materials). One aspect concerned the possible contribution of cells and NPs present during assay read-out to variability in the absorbance values, frequently causing them to exceed a value of 2.0, which is outside the linear dynamic range that is relevant to applying the Lambert-Beer law, as previously suggested by Xia et al. [33]. Therefore, in the revised SOP, laboratories were asked to read out the assay plates in the presence and absence of cells. Other critical steps in the test performance which cannot simply be addressed in the SOP, such as the verification of instruments or pipetting techniques, constituted major challenges to be faced in this field of research, to ensure quality of results. This led us to organize a focused training to resolve these issues. Additionally, the majority of participating laboratories indicated that they were not familiar with the principles of an inter-laboratory comparison, as exemplified by the submission of a large number of non-valid data. Training was thus held by well documented instructions and a teleconference to introduce the relevant principles of GLP and proficiency testing, to transfer the optimized SOP and quality criteria, and to discuss in detail good practices for enhancing the test performance. Subsequently, the final tier of the ILC study was launched.

3.3. Laboratories' Proficiency in Performing the In Vitro MTS Assay After Training (Tier 3)

In the third and final tier of the study, six laboratories were enrolled, of which five also had participated in the first phase (tier 1 and 2) with involvement of the same operator, and four had not been able to provide a full set of valid data in the second tier. The other laboratories dropped out mainly because of a shift or lack of resources at the time of the last tier. The participating laboratories were provided with fresh FBS and NPs, and used the optimized MTS assay SOP. All laboratories sent in data on three independent, valid runs performed on cells of different passage number from the same working cell bank vial. The dose-response data were evaluated for intra- and inter-laboratory biases, similar to the statistical analyses in tier 2 (Figures S5–S7). In Table 3, the mean ± SD values of EC30 and EC50 obtained by the six laboratories involved in the study before and after training are presented, together with the overall mean ± SD values of all laboratories.

Table 3. Comparison of intra- and inter-laboratory biases for EC30 and EC50 values before (tier 2) and after (tier 3) training in quality aspects of MTS assay performance. Individual data from the six laboratories that were trained, as well as the overall data of all laboratories (N; in bold and between brackets) are included. Mean and SD (and SD*, in bold and between brackets) values of calculated EC30 (nM staurosporine) and EC50 (μg/mL PS-NH2), coefficient of variation (CV) (%) and number of runs (n) are given. For tier 2, data from all (valid and non-valid) runs and valid runs only are indicated. Tier 3 data are presented in the presence and absence of cells. '-' indicates that no data were available, laboratory 6 participated only in tier 3.

		EC30 Staurosporine (nM)				**EC50 PS-NH2 (μg/mL)**			
	Laboratory	Mean	SD	CV (%)	Runs (n)	Mean	SD	CV (%)	Runs (n)
Before training (tier 2) All runs	1	539.9	63.7	11.8	2	83.7	12.1	14.4	8
	2	421.3	218.6	51.9	16	82.2	26.8	32.5	9
	3	275.0	79.4	28.9	7	75.6	9.2	12.2	4
	4	171.4	28.9	16.9	10	62.7	10.2	16.2	5
	5	129.4	2.4	1.8	4	69.4	10.5	15.1	2
	6	-	-	-	-	-	-	-	-
	All labs	**329.7 (N = 15)**	**182.0 (SD* = 78.7)**	**55.2**	-	**76.2 (N = 15)**	**20.4 (SD* = 10.6)**	**26.8**	-
Before training (tier 2) Valid runs	1	-	-	-	-	75.9	28.1	37.0	3
	2	303.4	159.9	52.7	6	-	-	-	-
	3	278.8	108.2	38.8	5	75.6	9.2	12.2	4
	4	160.1	81.8	51.1	5	-	-	-	-
	5	-	-	-	-	-	-	-	-
	6	-	-	-	-	-	-	-	-
	All labs	**261.0 (N = 8)**	**121.1 (SD* = 84.6)**	**46.4**	-	**76.2 (N = 4)**	**12.9 (SD* = 6.9)**	**16.9**	-
After training (tier 3) Cells present	1	78.1	18.3	23.4	6	59.5	7.6	12.8	3
	2	237.3	73.4	30.9	6	102.8	20.0	19.4	3
	3	156.2	23.6	15.1	6	71.3	3.9	5.4	3
	4	255.0	92.9	36.4	3	87.5	21.1	24.2	3
	5	290.4	40.4	13.9	6	87.5	9.5	10.9	3
	6	596.1	194.5	32.6	6	103.2	21.6	21.0	3
	All labs	**238.5 (N = 6)**	**118.3 (SD* = 59.4)**	**49.6**	-	**85.3 (N = 6)**	**19.7 (SD* = 16.5)**	**23.0**	-
After training (tier 3) Cells absent	1	82.9	11.1	13.4	6	56.9	4.7	8.2	3
	2	267.2	71.4	26.7	6	104.5	0.2	0.2	3
	3	162.0	21.5	13.3	6	70.5	4.8	6.9	3
	4	320.2	90.9	28.4	3	64.1	9.5	14.8	3
	5	284.6	29.4	10.3	6	85.3	8.9	10.4	3
	6	581.8	143.9	24.7	6	102.6	30.7	29.9	3
	All labs	**264.3 (N = 6)**	**140.0 (SD* = 53.1)**	**53.0**	-	**80.7 (N = 6)**	**22.8 (SD* = 7.7)**	**28.2**	-

When considering the entire data sets obtained in tier 2 (valid and non-valid runs) and tier 3 with read-out of the MTS assay in the presence of cells, the inter-laboratory variability assessed by means of CV (%) of all laboratories before (tier 2, N = 15) and after (tier 3, N = 6) training showed improvement due to training (respectively 55.2% compared to 49.6% for staurosporine, and 26.8% compared to 23.0% for PS-NH2). Furthermore, the reproducibility within the laboratories was found to be increased after training, which is obvious from the decrease in CV per laboratory (example of dose-response curves in Figure S8), as well as the decrease of overall SD* in the case of staurosporine (Table 3). Finally, reading out the MTS assay in the absence of cells resulted in a mean fitted EC30 of 264.3 ± 140.0 nM staurosporine and EC50 of 80.7 ± 22.8 μg/mL PS-NH2. Although this additional step did not further decrease the inter-laboratory variability (CV of all laboratories of 53.0% for staurosporine, and 28.2% for PS-NH2), it did further improve intra-laboratory performances for most laboratories as compared to cells present in the wells, evident again by decreased CV of individual laboratories and SD* (Figure S6 vs. Figure S5, Figure S7, and Table 3). Taking into account also the fact that in the second tier the majority of laboratories sent in non-valid run data (Table 1), because they were ignorant of good practice and ILC principles, these comparative analyses show that training improved the laboratories' proficiencies and ILC results.

4. Discussion

Nanotoxicological data reported in the literature have shown low reproducibility within and between different studies, and, consequently, conflicting conclusions. In the current ILC study, we aimed to tackle this problem by introducing SOPs for cytotoxicity testing via the MTS assay, using previously well characterized control NMs. In particular, we have evaluated the impact of training in these SOPs and in good laboratory practices on the proficiency of the participating laboratories.

The ILC study was composed of three tiers, including cell culturing and cell growth rate determination (tier 1), MTS assay for cytotoxicity measurement of NMs to evaluate the laboratories' inherent performance (tier 2) and, finally, a repetition of the MTS assay after training of the laboratories (tier 3). The process applied in this study has highlighted the degree of complexity that is related to good practice in nanosafety testing, where multiple consecutive steps of the workflow should all be tightly aligned, including material storage, NP dispersion, cell culture, cell seeding and exposure, and test performance. This has also been concluded in a round robin study performed by different partner laboratories of the German Priority Programme SPP1313, who observed that small variations in NP preparation, cell handling and the type of culture slide influenced NP stability and the outcomes of cell assays [57]. In another ILC study on the MTS cytotoxicity assay [34], system control measurements revealed similar steps in the protocol that are critical to ensure overall robustness and reproducibility of the assay results within and between laboratories. These factors have also been taken into account in ISO 19007:2018(E) [37].

In contrast to previous interlaboratory studies in the nanosafety research field, we here demonstrated that the availability of an optimized SOP in combination with training and the active implementation of good practice in test performance enhanced the quality of intra-laboratory test results, and improved inter-laboratory variability. The ILC study has also highlighted that principles of GLP, including the use of verified instruments and registration of each step in the execution, may provide guidance to enhance quality of results from in vitro toxicity assays. The developed SOPs for cell culture, cell growth rate determination and the MTS assay for the cytotoxicity assessment of NMs, were compiled to cover the different categories of test facility activities, including equipment, media and reagents, test and control items, the consecutive steps of the experimental protocol in a chronological order and data analysis. Furthermore, the SOPs included acceptance criteria to monitor test performance, forms to record the laboratory performance and observations, and calculation templates for the reporting of the test data. Each of these categories contained sufficient and explicit detail to ensure proper execution. Based on feedback collected from the participating laboratories through an online survey, however, we found that despite the availability of SOPs, deviations in the execution were frequently reported, for example, in the verification of laboratory instruments or the addition of antibiotics to the cell culture medium. Additionally, basic laboratory practices, such as pipetting techniques, which are not normally made explicit in SOPs were identified as a potential source of variability, in addition to influences from differences in technical infrastructure. This overall complexity accounts for the conflicting results on NPs' toxicity in the reported literature, and calls for increasing awareness and proper training of nanosafety professionals. Based on the feedback received after the second tier, the MTS assay SOP was revised to account for the identified hidden sources of variability, and additional training was provided to ensure all participating laboratories were familiar with the procedures and carefully followed all steps as detailed in the SOP. The results of the second ILC on the MTS assay showed that this process makes it possible to obtain higher reproducibility across independent laboratories from academia, research institutions and industry. The NIEHS Nano GO Consortium [33] has previously investigated NMs in several bioassays, including the MTS assay. They also found substantial variations in the results between different laboratories, and further stated that "frequent communication was very helpful for achieving reproducible results within and among the laboratories". However, no further details regarding the improvement of reproducibility were given. As similar issues related to the quality in test performance discussed in our study are also present among different operators within the same laboratory, it may be valuable to implement a similar process using shared materials as described here within laboratories, to determine and align the reproducibility in performance by multiple operators. A representative selection of positive and negative control materials, as well as SOPs with quality acceptance criteria and statistical data analysis, will make it possible to judge whether the data generated by multiple operators, even within a single laboratory, are reliable or not.

As a secondary outcome of this ILC study, we confirmed the suitability of PS-COOH and PS-NH2 as negative and positive control nanomaterials, respectively, to validate the test performance of the MTS assay. The PS-COOH NPs are complementary to other suggested negative nano-sized control materials for in vitro cell viability assays, such as barium sulfate [44] and carboxylated nanodiamonds [45]. Positively charged PS-NH2 NPs have already been included as validated positive control nanomaterials in the ISO 19007:2018(E) standard [37]. Based on the dose-response data obtained for the PS-NH2 in our large ILC study comprising a representative sample of research laboratories, an EC50 value of 80.7 ± 22.8 μg/mL was obtained which is exactly within the commonly accepted biological variability window of CV <30%. This EC50 for PS-NH2 was higher than the consensus EC50 value of 52.6 μg/mL (95% confidence intervals 44.1 to 62.6 μg/mL) for the same NPs in the A549 cell line reported by Elliott et al. [34], which can be attributed to differences in cell stocks, serum sources, cell seeding density and exposure times (24 vs. 48 h, respectively). Nevertheless, variability in the EC50 values in both studies was in a similar range and well below 30%, confirming the suitability of 50-nm PS-NH2 as positive control NPs in the MTS cytotoxicity assay. However, it should be mentioned that, in both studies, PS-NH2 from Bangs Laboratories Inc. have been used, which were not available anymore from the supplier at the time of this report. Therefore, to enable standardization in nanosafety testing, stable and effective reference NPs from a secured source are urgently needed.

We also stress that for this study, in order to focus—as a first step—on the sources of variability related to cell toxicity testing, we selected model PS NPs behaving well in terms of stability and dispersion, thus, many of the reproducibility issues highlighted in our study are not nano-specific. Reproducibility in in vitro nanotoxicity testing goes far beyond this simplistic view, as real nanoparticles, such as, for instance, metal oxides, introduce many more challenges related to their intrinsic (medium independent) and extrinsic (medium dependent) physical and chemical properties, which affect their dispersion and stability in cell culture medium, and—as a consequence of this—their fate and transport into cells, and, thus, the dose delivered to cells as a function of exposure time [58,59]. For example, surface affinity, which is dependent on particle and medium parameters, may cause the agglomeration or aggregation of nanomaterials, whereas particle size and density can affect the diffusion and sedimentation of NPs, thus affecting their in vitro exposure. Furthermore, especially for partially soluble NPs, such as metal and metal oxides, the rate of release of ions is depending on system parameters (e.g., pH, ligands present, flow conditions, etc.) and can greatly influence their in vitro toxicity potential. To allow for the correct interpretation of in vitro assay data using NPs closer to real applications, as opposed to model NPs with optimal dispersibility, standard protocols for the dispersion of nanomaterials in complex media, dispersion characterization and dosimetry, as described by Deloid et al. [59] and in ISO/TR 16196:2016 [60], should be adopted to ensure meaningful and reproducible quantification of in vitro delivered dose. The standardization of characterization methods to monitor the physicochemical properties before and after NPs' dispersion is also work in progress, e.g., by the OECD Series on the Safety of Manufactured Nanomaterials [58] and ISO/TC229 (Nanotechnologies), and has been subject of several ILC studies. Remarkably, in ILC studies in which model nanoparticles as polystyrene, gold or silica nanospheres have been used, for example those looking at size distribution measurements using nanoparticle tracking analysis [61], dynamic light scattering and centrifugal particle sedimentation [49], similar issues in reproducibility compared to our study have been reported, even for simple dispersions in buffer. They concluded that SOPs are indispensable for obtaining reliable and comparable NP size data, and should be tailored to the specific test system, sufficiently detailed and verified by a larger number of laboratories, to enable reproducible performance [49,61].

Other NP types than polystyrene have already been applied by others in the MTS assay, such as titanium dioxide, zinc oxide and multi-walled carbon nanotubes [33]. In addition to selecting the best protocol for NPs' dispersion and verification of the dispersion stability in complete CCM, as indicated above, these will require a case-by-case consideration of interference of the NPs with the optical density measurements. Although, initially, a number of laboratories observed optical interference of PS-NH2 with the MTS test read-out in our study (tier 2), we showed that, after good practice training and diligent application of the optimized SOP, this bias was resolved. This may be attributed to detailed instructions on NP dispersion in serum-containing medium and the use of the same serum batch by all laboratories. In addition, we introduced in the SOP a transfer of MTS medium to a new plate before optical read-out to avoid high absorbance contribution of the cells and NPs remaining in the wells. In this respect, our SOP deviates from the ILC study of Elliott et al. [34] and the ISO 19007:2018(E) standard [37], in which the average background absorbance level of the NP doses in culture medium are subtracted from each absorbance value of NP-exposed cells. In contrast, we recommend using an alternative cytotoxicity assay, based on a different optical read-out principle, in case NP interference is observed, as the correct assessment of the issue and adjustment is hampered by multiple factors, such as NP agglomeration, NP adherence to the cell surface or assay plate, cell-dependent NP uptake kinetics, etc.

Finally, it remains to be investigated whether the MTS assay SOP developed and tested by multiple laboratories in our study can be transferred to other cell types. Based on our experience, the SOP works optimally for adherent cell types growing in monolayers. The SOP can be adapted for use with suspension cells by introducing centrifugation steps to pellet the cells at the bottom of the wells before medium changes.

In conclusion, the experience reported in this study overall clearly indicates that the approach followed in this study, including inter-laboratory comparison studies using shared materials and detailed SOPs, together with training of the participants, can be used to optimize and generate robust SOPs, and to obtain reproducible data on NP cytotoxicity within and across independent laboratories.

Supplementary Materials: The following are available online at http://www.mdpi.com/2079-4991/10/8/1430/s1, Figure S1: Dosing Plate Layout for the MTS assay, Figure S2: Growth curve of the A549 cell line, Figure S3: Dose-response curves of cytotoxicity assessment using the MTS assay (Tier 2) from two individual laboratories, Figure S4: Intra- and inter-laboratory biases for percentage cell survival determined using the MTS assay based on all (valid and non-valid) runs of all laboratories (tier 2), Figure S5: Intra- and inter-laboratory biases for percentage cell survival determined using the MTS assay based on valid runs of laboratories after training in quality aspects of assay performance (tier 3), Figure S6: Intra- and inter-laboratory biases for percentage cell survival determined using the MTS assay based on valid runs of laboratories after training in quality aspects of assay performance (tier 3), Figure S7: Intra- and inter-laboratory biases for EC30 and EC50 values determined using the MTS assay based on valid runs of laboratories after training in quality aspects of assay performance (tier 3), Figure S8: Comparison of intra-laboratory bias in cytotoxicity assessment using the MTS assay before (tier 2) and after training in quality aspects of assay performance (tier 3), Standard Operating Procedure for A549 cell culturing, Standard Operating Procedure for assessment of A549 cell growth rate and viability, Calculation template for assessment of A549 cell growth rate and viability, Standard Operating Procedure for the MTS cytotoxicity assay, Calculation template for the MTS cytotoxicity assay.

Author Contributions: Conceptualization, I.N., A.H. (Andrea Haase), A.S. and K.A.D.; methodology, I.N., A.H. (Andrea Haase), S.A., A.J. and A.S.; coordination of ILC studies and training, I.N., A.H. (Andrea Haase) and A.S.; performance of ILC studies and data reporting, I.N., A.H. (Andrea Haase), S.A., L.R., A.J., C.R., A.L., J.-P.P., O.T., B.T., G.L., A.C.G., S.C., S.D., C.W., T.L.-F., Ã.G.-F., M.D., A.H. (Anna Huk), V.S., N.K., M.N., M.S., S.M. (Stefania Meschini), M.G.A., N.L., M.R., M.V., F.B., J.T., T.B., S.M. (Silvia Milani), J.R., A.S., K.A.D.; statistical data analysis and proficiency testing, H.W.; writing—original draft preparation, I.N., A.H. (Andrea Haase) and A.S.;

writing—review and editing, I.N., A.H. (Andrea Haase), S.A., L.R., A.J., H.W., C.R., A.L., J.-P.P., O.T., B.T., G.L., A.C.G., S.C., S.D., C.W., T.L.-F., Ã.G.-F., M.D., A.H. (Anna Huk), V.S., N.K., M.N., M.S., S.M. (Stefania Meschini), M.G.A., N.L., M.R., M.V., F.B., J.T., T.B., S.M. (Silvia Milani), J.R., A.S., K.A.D.; visualization of results, I.N., H.W.; funding acquisition, K.A.D. All authors have read and agreed to the published version of the manuscript.

Funding: This research received financial support from the EC FP7 project QualityNano (INFRA-2010-1.1.31-262163). The contribution of LIST was supported by the Fonds National de la Recherche of Luxembourg within the project NANION (FNR/12/SR/4009651). J.-P. Piret and O. Toussaint acknowledge the DG06 (Direction générale opérationnelle de l'Economie, de l'Emploi et de la Recherche) of the Walloon Region of Belgium for the project complement 'QNano' (2011-14) n°1117448. O. Toussaint, who unfortunately deceased in 2016, was a senior Research Associate from the Belgian F.R.S.-FNRS. The contribution of NILU was also supported by the NorNANoREG project (Norwegian Research Council, 239199/070). All authors are grateful to their institutions for supporting this work.

Acknowledgments: The authors would like to thank Siegfried Hofman (VITO) for advice concerning statistical methods for proficiency testing of laboratories and Christelle Saout (UNamur) for expert discussions on SOP development. We are grateful to Magda Marchetti (ISS), Pavel Rossner (IEM), Zuzana Novakova (IEM), Lisa Wollenberg (BfR), Dorit Mattern (KIT), Maximilian Kohnle (KIT), Kelly Blazy (INERIS) and Christelle Gamez (INERIS) for their excellent technical assistance.

Conflicts of Interest: The authors declare no conflict of interest.

Abbreviations

The following abbreviations are used in the manuscript:

CCM	Cell culture medium
CV	Coefficient of variation
DMSO	Dimethyl sulfoxide
EC30	Effective concentration causing 30% inhibition of cell viability
EC50	Effective concentration causing 50% inhibition of cell viability
FBS	Fetal bovine serum
GLP	Good laboratory practice
ILC	Inter-laboratory comparison
ISO	International Organization for Standardization
MTS	3-(4,5-dimethylthiazol-2-yl)-5-(3-carboxymethoxyphenyl)-2-(4-sulfophenyl)-2H-tetrazolium
MTT	3-(4,5-dimethylthiazol-2-yl)-2,5-diphenyltetrazolium bromide
NM	Nanomaterial
NP	Nanoparticle
OECD	Organization for Economic Cooperation and Development
PS-COOH	Carboxyl-modified polystyrene nanoparticles
PS-NH2	Amine-modified polystyrene nanoparticles
SD	Standard deviation
SOP	Standard operating procedure

References

1. Editorial, Join the dialogue. *Nat. Nanotechnol.* **2012**, *7*, 545. [CrossRef]
2. Editorial, The dialogue continues. *Nat. Nanotechnol.* **2013**, *8*, 69. [CrossRef] [PubMed]
3. Dawson, K.A. Leave the policing to others. *Nat. Nanotechnol.* **2013**, *8*, 73. [CrossRef] [PubMed]
4. Krug, H.F. Nanosafety research—Are we on the right track? *Angew. Chem. Int. Ed. Engl.* **2014**, *53*, 12304–12319. [CrossRef]
5. Ioannidis, J.P.A. Why most published research findings are false. *PLoS Med.* **2005**, *2*, e124. [CrossRef] [PubMed]
6. Begley, C.G.; Ellis, L.M. Drug development: Raise standards for preclinical cancer research. *Nature* **2012**, *483*, 531–533. [CrossRef] [PubMed]
7. McNutt, M. Journals unite for reproducibility. *Science* **2014**, *346*, 679. [CrossRef] [PubMed]
8. Prinz, F.; Schlange, T.; Asadullah, K. Believe it or not: How much can we rely on published data on potential drug targets? *Nat. Rev. Drug Discov.* **2011**, *10*, 712. [CrossRef]
9. Berry, C. Reproducibility in experimentation—The implications for regulatory toxicology. *Toxicol. Res.* **2014**, *3*, 411–417. [CrossRef]

10. Alberts, B.; Cicerone, R.J.; Fienberg, S.E.; Kamb, A.; McNutt, M.; Nerem, R.M.; Schekman, R.; Shiffrin, R.; Stodden, V.; Suresh, S.; et al. Scientific integrity. Self-correction in science at work. *Science* **2015**, *348*, 1420–1422. [CrossRef]
11. Schneider, K. Faking it: The case against Industrial Bio-Test Laboratories. *Amicus J.* **1983**, 14–26.
12. Cox, C. Glyphosate, Part 1: Toxicology. *J. Pesticide Reform* **1995**, *15*. Available online: http://www.1hope.org/glyphos8.htm (accessed on 20 July 2020).
13. Seiler, J.P. *Good Laboratory Practice—The Why and the How*, 2nd ed.; Springer: Heidelberg, Germany, 2005. [CrossRef]
14. Hristozov, D.R.; Gottardo, S.; Critto, A.; Marcomini, A. Risk assessment of engineered nanomaterials: A review of available data and approaches from a regulatory perspective. *Nanotoxicology* **2012**, *6*, 880–898. [CrossRef] [PubMed]
15. DaNa 2.0 Knowledge Base Nanomaterials—Methodology for Selection of Publications (Version 2016). Available online: http://www.nanoobjects.info/en/nanoinfo/methods/991-literature-criteria-checklist (accessed on 22 June 2020).
16. Vankoningsloo, S.; Piret, J.P.; Saout, C.; Noel, F.; Mejia, J.; Zouboulis, C.C.; Delhalle, J.; Lucas, S.; Toussaint, O. Cytotoxicity of multi-walled carbon nanotubes in three skin cellular models: Effects of sonication, dispersive agents and corneous layer of reconstructed epidermis. *Nanotoxicology* **2010**, *4*, 84–97. [CrossRef] [PubMed]
17. Maiorano, G.; Sabella, S.; Sorce, B.; Brunetti, V.; Malvindi, M.A.; Cingolani, R.; Pompa, P.P. Effects of cell culture media on the dynamic formation of protein-nanoparticle complexes and influence on the cellular response. *ACS Nano* **2010**, *4*, 7481–7491. [CrossRef]
18. Taurozzi, J.S.; Hackley, V.A.; Wiesner, M.R. Ultrasonic dispersion of nanoparticles for environmental, health and safety assessment—Issues and recommendations. *Nanotoxicology* **2011**, *5*, 711–729. [CrossRef]
19. Monopoli, M.P.; Aberg, C.; Salvati, A.; Dawson, K.A. Biomolecular coronas provide the biological identity of nanosized materials. *Nat. Nanotechnol.* **2012**, *7*, 779–786. [CrossRef]
20. Drescher, D.; Orts-Gil, G.; Laube, G.; Natte, K.; Veh, R.W.; Osterle, W.; Kneipp, J. Toxicity of amorphous silica nanoparticles on eukryotic cell model is determined by particle agglomeration and serum protein adsorption effects. *Anal. Bioanal. Chem.* **2011**, *400*, 1367–1373. [CrossRef]
21. Lesniak, A.; Fenaroli, F.; Monopoli, M.P.; Åberg, C.; Dawson, K.A.; Salvati, A. Effects of the presence or absence of a protein corona on silica nanoparticle uptake and impact on cells. *ACS Nano* **2012**, *6*, 5845–5857. [CrossRef]
22. Wang, F.; Yu, L.; Monopoli, M.P.; Sandin, P.; Mahon, E.; Salvati, A.; Dawson, K.A. The biomolecular corona is retained during nanoparticle uptake and protects the cells from the damage induced by cationic nanoparticles until degraded in the lysosomes. *Nanomedicine* **2013**, *9*, 1159–1168. [CrossRef]
23. Ge, C.; Du, J.; Zhao, L.; Wang, L.; Liu, Y.; Li, D.; Yang, Y.; Zhou, R.; Zhao, Y.; Chai, Z.; et al. Binding of blood proteins to carbon nanotubes reduces cytotoxicity. *Proc. Natl. Acad. Sci. USA* **2011**, *108*, 16968–16973. [CrossRef] [PubMed]
24. Hu, W.; Peng, C.; Lv, M.; Li, X.; Zhang, Y.; Chen, N.; Fan, C.; Huang, Q. Protein corona-mediated mitigation of cytotoxicity of graphene oxide. *ACS Nano* **2011**, *5*, 3693–3700. [CrossRef] [PubMed]
25. Salvati, A.; Pitek, A.S.; Monopoli, M.P.; Prapainop, K.; Bombelli, F.B.; Hristov, D.R.; Kelly, P.M.; Åberg, C.; Mahon, E.; Dawson, K.A. Transferrin-functionalized nanoparticles lose their targeting capabilities when a biomolecule corona adsorbs on the surface. *Nat. Nanotechnol.* **2013**, *8*, 137–143. [CrossRef] [PubMed]
26. Kim, J.A.; Salvati, A.; Åberg, C.; Dawson, K.A. Suppression of nanoparticle cytotoxicity approaching in vivo serum concentrations: Limitations of in vitro testing for nanosafety. *Nanoscale* **2014**, *6*, 14180–14184. [CrossRef]
27. Francia, V.; Yang, K.; Deville, S.; Reker-Smit, C.; Nelissen, I.; Salvati, A. Corona Composition Can Affect the Mechanisms Cells Use to Internalize Nanoparticles. *ACS Nano* **2019**, *13*, 11107–11121. [CrossRef]
28. Guadagnini, R.; Halamoda Kenzaoui, B.; Cartwright, L.; Pojana, G.; Magdolenova, Z.; Bilanicova, D.; Saunders, M.; Juillerat, L.; Marcomini, A.; Huk, A.; et al. Toxicity screenings of nanomaterials: Challenges due to interference with assay processes and components of classic in vitro tests. *Nanotoxicology* **2015**, *9*, 13–24. [CrossRef]
29. Yu, M.; Selvaraj, S.K.; Liang-Chu, M.M.; Aghajani, S.; Busse, M.; Yuan, J.; Lee, G.; Peale, F.; Klijn, C.; Bourgon, R.; et al. A resource for cell line authentication, annotation and quality control. *Nature* **2015**, *520*, 307–311. [CrossRef]

30. Nübling, C.M.; Baylis, S.A.; Hanschmann, K.M.; Montag-Lessing, T.; Chudy, M.; Kress, J.; Ulrych, U.; Czurda, S.; Rosengarten, R.; Mycoplasma Collaborative Study Group. World Health Organization International Standard To Harmonize Assays for Detection of Mycoplasma DNA. *Appl. Environ. Microbiol.* **2015**, *81*, 5694–5702. [CrossRef]
31. Lundholt, B.K.; Scudder, K.M.; Pagliaro, L. A simple technique for reducing edge effect in cell-based assays. *J. Biomol. Screen* **2003**, *8*, 566–570. [CrossRef]
32. Mosmann, T. Rapid colorimetric assay for cellular growth and survival: Application to proliferation and cytotoxicity assays. *J. Immunol. Methods* **1983**, *65*, 55–63. [CrossRef]
33. Xia, T.; Hamilton, R.F.; Bonner, J.C.; Crandall, E.D.; Elder, A.; Fazlollahi, F.; Girtsman, T.A.; Kim, K.; Mitra, S.; Ntim, S.A.; et al. Inter-laboratory evaluation of in vitro cytotoxicity and inflammatory responses to engineered nanomaterials: The NIEHS Nano GO Consortium. *Environ. Health Perspect.* **2013**, *121*, 683–690. [CrossRef]
34. Elliott, J.T.; Rösslein, M.; Song, N.W.; Toman, B.; Kinsner-Ovaskainen, A.; Maniratanachote, R.; Salit, M.L.; Petersen, E.J.; Seqeira, F.; Romsos, E.L.; et al. Toward achieving harmonization in a nanocytotoxicity assay measurement through an interlaboratory comparison study. *ALTEX* **2017**, *34*, 201–218. [CrossRef] [PubMed]
35. Piret, J.-P.; Bondarenko, O.M.; Boyles, M.S.P.; Himly, M.; Ribeiro, A.R.; Benetti, F.; Smal, C.; Lima, B.; Potthoff, A.; Simion, M.; et al. Pan-European Inter-Laboratory Studies on a Panel of in Vitro Cytotoxicity and Pro-Inflammation Assays for Nanoparticles. *Arch. Toxicol.* **2017**, *91*, 2315–2330. [CrossRef] [PubMed]
36. Rösslein, M.; Elliott, J.T.; Salit, M.; Petersen, E.J.; Hirsch, C.; Krug, H.F.; Wick, P. Use of Cause-and-Effect Analysis to Design a High-Quality Nanocytotoxicology Assay. *Chem. Res. Toxicol.* **2015**, *28*, 21–30. [CrossRef] [PubMed]
37. International Standard ISO 19007:2018(E). *Nanotechnologies—In Vitro Mts Assay for Measuring the Cytotoxic Effect of Nanoparticles*; ISO: Geneva, Switzerland, 2018.
38. International Standard ISO/IEC 17043:2010. *Conformity Assessment—General Requirements for Proficiency Testing*; ISO: Geneva, Switzerland, 2010.
39. Monteiro-Riviere, N.A.; Inman, A.O.; Zhang, L.W. Limitations and relative utility of screening assays to assess engineered nanoparticle toxicity in a human cell line. *Toxicol. Appl. Pharmacol.* **2009**, *234*, 222–235. [CrossRef]
40. International Standard ISO 10993-5:2009. *Biological Evaluation of Medical Devices—Part 5: Tests for In Vitro Cytotoxicity*; ISO: Geneva, Switzerland, 2009.
41. Gangwal, S.; Brown, J.S.; Wang, A.; Houck, K.A.; Dix, D.J.; Kavlock, R.J.; Hubal, E.A. Informing selection of nanomaterial concentrations for ToxCast in vitro testing based on occupational exposure potential. *Environ. Health Perspect.* **2011**, *119*, 1539–1546. [CrossRef]
42. Panas, A.; Marquardt, C.; Nalcaci, O.; Bockhorn, H.; Baumann, W.; Paur, H.R.; Mülhopt, S.; Diabaté, S.; Weiss, C. Screening of different metal oxide nanoparticles reveals selective toxicity and inflammatory potential of silica nanoparticles in lung epithelial cells and macrophages. *Nanotoxicology* **2013**, *7*, 259–273. [CrossRef]
43. Moche, H.; Chevalier, D.; Barois, N.; Lorge, E.; Claude, N.; Nesslany, F. Tungsten Carbide-Cobalt as a Nanoparticulate Reference Positive Control in in Vitro Genotoxicity Assays. *Toxicol. Sci.* **2014**, *137*, 125–134. [CrossRef]
44. Loza, K.; Föhring, I.; Bünger, J.; Westphal, G.A.; Köller, M.; Epple, M.; Sengstock, C. Barium sulfate micro- and nanoparticles as bioinert reference material in particle toxicology. *Nanotoxicology* **2016**, *10*, 1492–1502. [CrossRef]
45. Paget, V.; Sergent, J.A.; Grall, R.; Altmeyer-Morel, S.; Girard, H.A.; Petit, T.; Gesset, C.; Mermoux, M.; Bergonzo, P.; Arnault, J.C.; et al. Carboxylated Nanodiamonds Are Neither Cytotoxic Nor Genotoxic on Liver, Kidney, Intestine and Lung Human Cell Lines. *Nanotoxicology* **2014**, *8*, 46–56. [CrossRef]
46. Mülhopt, S.; Diabaté, S.; Dilger, M.; Adelhelm, C.; Anderlohr, C.; Bergfeldt, T.; Gómez de la Torre, J.; Jiang, Y.; Valsami-Jones, E.; Langevin, D.; et al. Characterization of Nanoparticle Batch-To-Batch Variability. *Nanomaterials* **2018**, *8*, 311. [CrossRef] [PubMed]
47. Kim, J.A.; Åberg, C.; Salvati, A.; Dawson, K.A. Role of cell cycle on the cellular uptake and dilution of nanoparticles in a cell population. *Nat. Nanotechnol.* **2012**, *7*, 62–68. [CrossRef] [PubMed]
48. Wang, F.; Bexiga, M.G.; Anguissola, S.; Boya, P.; Simpson, J.C.; Salvati, A.; Dawson, K.A. Time resolved study of cell death mechanisms induced by amine-modified polystyrene nanoparticles. *Nanoscale* **2013**, *5*, 10868–10876. [CrossRef] [PubMed]

49. Langevin, D.; Lozano, O.; Salvati, A.; Kestens, V.; Monopoli, M.; Raspaud, E.; Mariot, S.; Salonen, A.; Thomas, S.; Driessen, M.; et al. Inter-laboratory comparison of nanoparticle size measurements uisng dynamic light scattering and differential centrifugal sedimentation. *NanoImpact* **2018**, *10*, 97–107. [CrossRef]
50. Bexiga, M.G.; Varela, J.A.; Wang, F.; Fenaroli, F.; Salvati, A.; Lynch, I.; Simpson, J.C.; Dawson, K.A. Cationic nanoparticles induce caspase 3-, 7- and 9-mediated cytotoxicity in a human astrocytoma cell line. *Nanotoxicology* **2011**, *5*, 557–567. [CrossRef]
51. Ruenraroengsak, P.; Novak, P.; Berhanu, D.; Thorley, A.J.; Valsami-Jones, E.; Gorelik, J.; Korchev, Y.E.; Tetley, T.D. Respiratory epithelial cytotoxicity and membrane damage (holes) caused by amine-modified nanoparticles. *Nanotoxicology* **2012**, *6*, 94–108. [CrossRef]
52. Deville, S.; Honrath, B.; Tran, Q.T.D.; Fejer, G.; Lambrichts, I.; Nelissen, I.; Dolga, A.M.; Salvati, A. Time-resolved characterization of the mechanisms of toxicity induced by silica and amino-modified polystyrene on alveolar-like macrophages. *Arch. Toxicol.* **2020**, *94*, 173–186. [CrossRef]
53. Bal-Price, A.; Coecke, S. Guidance on Good Cell Culture Practice (GCCP). In *Cell Culture Techniques, Neuromethods*; Aschner, M., Suñol, C., Bal-Price, A., Eds.; Humana Press: Totowa, NJ, USA, 2011; Volume 56, pp. 1–25.
54. R Core Team, R: A Language and Environment for Statistical Computing. R Foundation for Statistical Computing, Vienna, Austria. Available online: https://www.R-project.org/ (accessed on 1 October 2015).
55. Ritz, C.; Streibig, J.C. Bioassay Analysis using R. *J. Statist. Softw.* **2005**, *12*, 1–22. [CrossRef]
56. International Standard ISO 13258:2005(E). *Statistical Methods for Use in Proficiency Testing by Inter-Laboratory Comparisons*; ISO: Geneva, Switzerland, 2005.
57. Landgraf, L.; Nordmeyer, D.; Schmiel, P.; Gao, Q.; Ritz, S.; Gebauer, S.; Graß, S.; Diabaté, S.; Treuel, L.; Graf, C.; et al. Validation of weak biological effects by round robin experiments: Cytotoxicity/biocompatibility of SiO2 and polymer nanoparticles in HepG2 cells. *Sci. Rep.* **2017**, *7*, 4341. [CrossRef]
58. Gao, X.; Lowry, G.V. Progress towards standardized and validated characterizations for measuring physicochemical properties of manufactured nanomaterials relevant to nano health and safety risks. *NanoImpact* **2018**, *9*, 14–30. [CrossRef]
59. DeLoid, G.M.; Cohen, J.M.; Pyrgiotakis, G.; Demokritou, P. An integrated dispersion preparation, characterization and in vitro dosimetry methodology for engineered nanomaterials. *Nat. Protoc.* **2017**, *12*, 355–371. [CrossRef] [PubMed]
60. International Standard ISO/TR 16196:2016. *Nanotechnologies—Compilation and Description of Sample Preparation and Dosing Methods for Engineered and Manufactured Nanomaterials*; ISO: Geneva, Switzerland, 2016.
61. Hole, P.; Sillence, K.; Hannell, C.; Maguire, C.M.; Roesslein, M.; Suarez, G.; Capracotta, S.; Magdolenova, Z.; Horev-Azaria, L.; Dybowska, A.; et al. Interlaboratory comparison of size measurements on nanoparticles using nanoparticle tracking analysis (NTA). *J. Nanopart. Res.* **2013**, *15*, 2101. [CrossRef]

Article

Attachment Efficiency of Nanomaterials to Algae as an Important Criterion for Ecotoxicity and Grouping

Kerstin Hund-Rinke [1,*], Tim Sinram [1], Karsten Schlich [1], Carmen Nickel [2], Hanna Paula Dickehut [3], Matthias Schmidt [3] and Dana Kühnel [3]

1 Fraunhofer Institute for Molecular Biology and Applied Ecology, Auf dem Aberg 1, 57392 Schmallenberg, Germany; t.sinram@googlemail.com (T.S.); karsten.schlich@ime.fraunhofer.de (K.S.)

2 Institute for Energy and Environmental Technology, e.V. (IUTA), Bliersheimer Straße 58-60, 47229 Duisburg, Germany; nickel@iuta.de

3 Helmholtz Centre for Environmental Research—UFZ, Permoserstr. 15, 04318 Leipzig, Germany; Hanna.P.D@gmx.de (H.P.D.); matthias.schmidt@ufz.de (M.S.); dana.kuehnel@ufz.de (D.K.)

* Correspondence: kerstin.hund-rinke@ime.fraunhofer.de

Received: 6 March 2020; Accepted: 20 May 2020; Published: 27 May 2020

Abstract: Engineered nanomaterials (ENMs) based on CeO_2 and TiO_2 differ in their effects on the unicellular green alga *Raphidocelis subcapitata* but these effects do not reflect the physicochemical parameters that characterize such materials in water and other test media. To determine whether interactions with algae can predict the ecotoxicity of ENMs, we studied the attachment of model compounds (three subtypes of CeO_2 and five subtypes of TiO_2) to algal cells by light microscopy and electron microscopy. We correlated our observations with EC_{50} values determined in growth inhibition assays carried out according to the Organisation for Economic Co-operation and Development (OECD) test guideline 201. Light microscopy revealed distinct patterns of ENM attachment to algal cells according to the type of compound, with stronger interactions leading to greater toxicity. This was confirmed by electron microscopy, which allowed the quantitative assessment of particle attachment. Our results indicate that algal extracellular polymeric substances play an important role in the attachment of ENMs, influencing the formation of agglomerates. The attachment parameters in short-term tests predicted the toxicity of CeO_2 and TiO_2 ENMs and can be considered as a valuable tool for the identification of sets of similar nanoforms as requested by the European Chemicals Agency in the context of grouping and read-across.

Keywords: nanotoxicology; European Chemicals Agency (ECHA); ecotoxicology; nanoparticles; aggregation; *Raphidocelis subcapitata*

1. Introduction

Engineered nanomaterials (ENMs) show great variation in size, shape, crystalline structure, and surface modifications. According to the European Chemicals Agency (ECHA), grouping and read-across approaches can be applied to reduce the number of tests required for the risk assessment of ENMs [1]. ENM groups with analogous sets of physicochemical properties enable reasonable hazard predictions without additional testing, thus saving time and costs. Most concepts for the prediction of ENM properties focus on toxicity in humans [2,3]. Insight into ecotoxicity and grouping has been gained in systematic studies that generated ecotoxicological data for seven chemical species (Ag, ZnO, TiO_2, CeO_2, Cu, Fe, and SiO_2) with 25 modifications [4,5]. Given the focus on regulatory applications, ecotoxicity was based on the Organisation for Economic Co-operation and Development (OECD) test guidelines 201 (algae), 202 (daphnids), and 236 (fish embryos). The studies considered reactivity, ion release, and morphology as properties indicating ecotoxicity. Nevertheless, it was difficult to separate subtypes of the same chemical species with this grouping approach. The ecotoxicity of TiO_2 and CeO_2 showed particularly broad ranges of subtype-dependent EC_{50} values (TiO_2 = 0.2–126.9 mg/L, CeO_2 = 8.5–99 mg/L). This suggests that additional parameters are needed to improve grouping, such as the adsorption of ENMs to algae [5]. The attachment of nanomaterials to green algae has already been reported [6–9] but these studies have not systematically addressed the attachment of different subtypes of the same ENM (and the relationship with ecotoxicity) or the quantity of ENMs attached to the algal cells.

We therefore investigated the attachment of ENMs to algae in order to determine whether this parameter can improve the results of ecotoxicological grouping. We focused on ENMs based on three subtypes of CeO_2 and five subtypes of TiO_2 differing in their ecotoxicological impact on algae, representing a subset of nanomaterials that have been comprehensively tested for aquatic and terrestrial ecotoxicity [4,5]. We studied the interaction between the ENMs and algal cells by light microscopy and quantified their behavior by scanning electron microscopy (SEM) with energy-dispersive X-ray spectroscopy (EDX).

2. Materials and Methods

2.1. Selection of Nanomaterials

The eight ENMs (three subtypes of CeO_2 and five subtypes of TiO_2) have been characterized in detail, and their physicochemical properties (and the corresponding analytical methods) are described in the supporting information of two publications [4,5]. The ENMs considered herein originated mainly from the program "Testing a Representative set of Manufactured Nanomaterials" initiated by the OECD Working Party on Manufactured Nanomaterials [10,11]. Selected characteristics of the ENMs are listed in Table 1. Each ENM was used at a concentration of 100 mg/L to determine the agglomerate size, zeta-potential, and reactivity.

Table 1. Selected physicochemical characteristics of the eight engineered nanomaterials (ENMs) investigated in this study (for further details see the supporting information in two earlier publications [4,5]).

Nanomaterial	Primary Particle Size (SEM/TEM) (nm)	Surface Area (m^2/g) [1]	Agglomerate Size–Z Average (nm) (DLS) [2]	Zeta-Potential (mV) [2]	Reactivity (DMPO) [2,3]	Crystalline Structure	Coating
CeO_2 NM-211	4–15	66	442 ± 85	−19.8	0.81	Cubic cereonite	Uncoated
CeO_2 NM-212	40	27	831 ± 209	−20.4	0.96	Cubic cereonite	Uncoated
CeO_2 NM-213	35	4	1042 ± 178	−25.9	1.1	Cubic cereonite	Uncoated
TiO_2 NM-104	30	60	1596 ± 498	−0.9	1.1 UV activation: 1.6	Rutile	Al_2O_3 (6%) coating and glycerol (1%) functionalization
TiO_2 NM-105	21	51	1409 ± 533	−2.4	1.0 UV activation: 20.8	14% Rutile 86% anatase	Uncoated
TiO_2 Eu-doped	19 (BET)	148	1612 ± 384	−23.1	0.7 UV activation: 1.4	Mainly rutile	Uncoated
TiO_2 Fe-doped	10 (BET)	63	1866 ± 106	−21.3	1.0 UV activation: 1.4	Mainly rutile	Uncoated
TiO_2 non-doped	15	78	743 ± 859	−22	0.7 UV activation: 1.5	9% Rutile, 91% anatase	Uncoated

[1] Based on the BET (information provided by the manufacturers). [2] Determined in the Organisation for Economic Co-operation and Development (OECD) test medium [12] used for the growth test with algae. The pH was not adjusted but reached values between 7.0 and 7.4 (values >1.3 indicate reactivity). The values are presented as sample-to-blank ratios (n = 3).

[3] Measurement of hydroxyl radicals generated after UV irradiation via Fenton-type reactions in the presence of hydrogen peroxide and 5,5-dimethyl-1-pyrroline-*N*-oxide (DMPO) [13,14]. SEM = scanning electron microscopy, TEM = transmission electron microscopy, DLS = dynamic light scattering, BET = Brunauer, Emmett and Teller specific surface area, UV = ultraviolet.

2.2. Preparation of Suspensions

The ENM suspensions were prepared as previously described [4]. Briefly, a stock suspension of each ENM (1 mg/mL) was prepared in ultrapure water by sonicating for 10 min using a cup horn (Bandelin, Germany) with a final energy input of 0.6 W/mL. A specific amount of the stock suspension was then applied to the test medium to achieve the target concentration for subsequent tests.

2.3. Algal Growth Inhibition Test

Growth of the green alga *Raphidocelis subcapitata* was measured as set out in the OECD test guideline 201 [12]. The growth rate was calculated by measuring chlorophyll fluorescence in vitro [15], with four replicates per test concentration and eight replicates for the control. Five to six test concentrations with a spacing factor of 2–3 were prepared. Furthermore, growth inhibition was compared in high-ionic-strength Grimme–Broadman (GB) medium and OECD medium (Table 2).

Table 2. Composition of Grimme–Broadman (GB) medium [16] and OECD medium [12].

	GB Medium (µmol/L)	OECD Medium (µmol/L)
KNO_3	8000	–
NaCl	8000	–
$MgSO_4 * 7\,H_2O$	1000	60.9
$Na_2HPO_4 * 2\,H_2O$	1000	–
$NaH_2PO_4 * H_2O$	3000	–
$CaCl_2 * 2\,H_2O$	100	122
$MnCl_2 * 4\,H_2O$	2.5	2.1
H_3BO_3	8	2.99
$ZnSO_4 * 7\,H_2O$	0.7	–
$Na_2MoO_4 * 2\,H_2O$	0.016	0.0289
FeEDTA [1]	25	–
NH_4Cl	–	280
KH_2PO_4	–	9.19
$MgCl_2 * 6\,H_2O$	–	59
$ZnCl_2$	–	0.022
$CoCl_2 * 6\,H_2O$	–	0.0063
$CuCl_2 * 2\,H_2O$	–	0.00006
$Na_2EDTA * 2\,H_2O$	–	0.269
$FeCl_3 * 6\,H_2O$	–	0.237
Ionic strength	24.349	1.602

[1] Composed of $FeSO_3$; $* 7\,H_2O$ and $Na_2EDTA * 2\,H_2O$.

2.4. Microscopy

Particle attachment to algae was observed by light and electron microscopy, the former for the rapid and inexpensive screening of the ENMs and the latter for more detailed tests at the single-cell level. Image evaluation was then used to estimate the coverage of cells by ENM particles, but this was labor-intensive and only a few individual cells could be analyzed per sample, reducing the statistical power. Image analysis was also unable to account for particles attached underneath the cells. The advantages and disadvantages of each method are summarized in Table 3.

2.4.1. Attachment of ENMs to Algae by Light Microscopy

We carried out a short-term test and a growth inhibition test, and in each case observed the algae by light microscopy to investigate their interactions with the ENMs. For the short-term test, an algal culture was incubated in OECD medium (Table 2) until the cell density reached 3–4 million cells/mL. We then transferred 90 mL of this culture to a clean, sterile 250-mL Schott Duran Erlenmeyer flask and added 10 mL of the ENM stock dispersion to achieve a final concentration of 100 mg/L. The flask was then incubated for 3 h under the same conditions as the algal growth inhibition tests before removing samples for analysis. In addition, 3-h spike tests were carried out with fixed concentrations of NM-212

(100 mg/L) and varying algal cell densities (3,000,000, 1,400,000, 700,000 and 175,000 cells/mL). For microscopic analysis during the growth test, algae incubated with selected ENM concentrations were analyzed at the test end.

Table 3. Advantages and disadvantages of light and electron microscopy for the observation of particle–cell interactions.

Light Microscopy	Electron Microscopy
Time and cost efficient	Time and cost intensive
No fixation and preparation	Fixation and preparation of samples may lead to artifacts (e.g., particles may be washed off)
Qualitative	Quantification of attachment possible
No identification of elements (may lead to artifacts)	Element identification ensures particle identity
Lower resolution compared to electron microscopy	Lower concentrations of particles and smaller particles / agglomerates can be detected

We pipetted 20–50 µL of algal culture (growth tests and spike tests) onto a clean microscope slide and placed a cover glass on top. The slide was air dried at room temperature until the liquid under the cover glass had partially evaporated. We then sealed the edges to prevent further drying (which would cause the cells to shrivel and prevent detailed observation). To avoid the influence of physical or chemical parameters on the attached nanoparticles, we avoided reagents and protocols that might affect the structural integrity of the sample (e.g., paraformaldehyde or heat fixation). The algae were then observed using a Leica Primo Star (Leica Microsystems, Wetzlar, Germany) equipped with three Plan Achromat oil immersion objectives (10×, 40×, and 100×) as well as filters for phase-contrast microscopy. Only objects in the aqueous phase were considered. All images were captured using an AxioCam Erc 5s with an additional 10× lens (Carl Zeiss, Jena, Germany) for final magnifications of 100×, 400×, and 1000×.

2.4.2. Attachment of ENMs to Algae by Electron Microscopy (SEM-EDX)

The algal cells were exposed to 0.1 or 1 mg/L CeO_2 NM-212 for 24 h in OECD or GB medium as described above for the growth inhibition test. For each concentration, we collected 5–6 samples of 10 mL. Each algal–ENM suspension was centrifuged for 5 min at 4400× *g* in a Heraeus Megafuge (Heraeus Institute, Hanau, Germany). The supernatant was discarded, and the pellet gently resuspended in 1 mL 4% paraformaldehyde in sodium cacodylate buffer, pH 7.4 (Electron Microscopy Sciences, Hatfield, PA, USA) for 15 h at 4 °C. The cells were then deposited onto polycarbonate filters (0.2 µm pore size) and washed for 15 min in sodium cacodylate buffer, before gradual dehydration (transfer to 30% ethanol in MilliQ water followed by the dropwise addition of 96% ethanol over 30 min to increase the final ethanol concentration to 93%). The filters were then transferred to absolute ethanol and critical point dried using a Leica EM CPD 300 device. In preparation for SEM, the samples were sputter-coated with a 30-nm gold–palladium (90/10) layer using a Leica EM SCD 500 instrument and mounted onto SEM stubs.

Samples were observed under a Zeiss Merlin VP Compact field-emitting SEM (Carl Zeiss Microscopy, Oberkochen, Germany) with a Bruker ((Bruker Nano Analytics, Berlin, Germany)) QUANTAX FlatQUAD X-ray spectrometer. The electron acceleration voltage was set to 10 kV and the beam current to ~300 pA, which achieved the ionization of cerium while allowing for cell surface imaging with an Everhard-Thornley secondary electron detector. Cerium maps were obtained from spatially resolved EDX data using the Ce L-alpha line.

The coverage of algal cells with ENMs was determined from overlays of SEM and EDX images generated using the Correlia Plugin for ImageJ/Fiji [17,18]. Based on these overlays, Fiji tools were used to calculate the area of each image covered with algae and the area covered with cerium as well as algae.

From these values, we were able to calculate the coverage as a percentage. In each of the 5–6 samples collected from the two different media (GB and OCED) and the two different ENM concentrations (0.1 and 1 mg/L), we analyzed 2–7 individual cells by EDX in order to determine the mean coverage.

2.5. *Statistical Evaluation*

Statistical analysis and the calculation of EC_{50} values in the algal growth inhibition tests were carried out using ToxRat Professional v3.3 (ToxRat Solutions, Alsdorf, Germany). Dose-response functions were determined by linear regression (probit model) and EC_{50} confidence limits were based on Fieller's theorem.

3. Results

3.1. *Attachment of* CeO_2

3.1.1. Light Microscopy

CeO_2 NM-212 showed strong attachment to algal cells during spike experiments after a contact period of 3 h (Figures 1 and 2). There was mostly no direct contact with the cell wall, but instead the ENMs attached to a transparent sheath around the individual algal cells. We also observed the attachment of NM-212 to a transparent structure, with a shape similar to the algal cells. The formation of agglomerates was highly dependent on the concentration ratio between the algal cells and nanoparticles. The 3-h spike experiments showed that small agglomerates (~0.1 mm) formed at the highest and lowest cell densities (3,000,000 and 175,000 cells/mL), whereas larger agglomerates (0.5–2 mm) formed at the intermediate cell densities (1,400,000 and 700,000 cells/mL). Although the appearance of the agglomerates was dependent on cell density, the attachment of ENMs to individual algal cells was similar across all spike experiments. Even algal cells embedded within thicker and larger agglomerates featured the same gap between the cell surface and attached nanoparticles. These transparent structures were not observed in the controls without NM-212, indicating they were induced by the presence of the ENM.

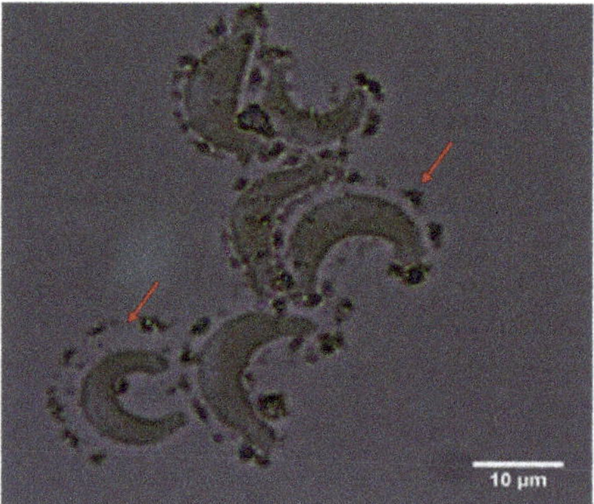

Figure 1. Nanoparticle attachment to transparent, sheath-like structures around algal cells (red arrows). This phase-contrast image (1000× magnification) was captured after 3 h incubation with 100 mg/L CeO_2 NM-212.

The attachment of ENMs to the algal cells was observed in the algal growth inhibition tests at all test concentrations after 72 h. As described for the spike experiments, the size of the agglomerates was highly dependent on the concentration ratio between the algal cells and nanoparticles (Figures S1–S3). The size of agglomerates, as determined by microscopy, initially increased in line with the ENM test concentration (2.5–10 mg/L) but decreased again at the highest value (40 mg/L). The distinct shell-like attachment, which we identified during spike experiments, was less pronounced during the growth inhibition tests perhaps due to the lower ENM concentrations. CeO_2 NM-211 showed attachment behavior comparable to CeO_2 NM-212 (Figure S4).

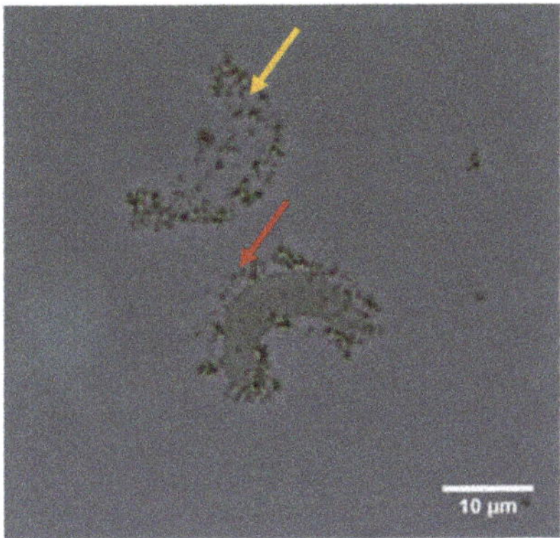

Figure 2. Nanoparticle attachment to transparent, algal-shaped structures with (red arrow) and without (yellow arrow) a corresponding algal cell. This phase-contrast image (1000× magnification) was captured after 3 h incubation with 100 mg/L CeO_2 NM-212.

In contrast, CeO_2 NM-213 showed much weaker attachment to *R. subcapitata* cells during both the spike experiments and algal growth inhibition tests. Despite this weak interaction, a transparent sheath-like structure was again observed around the algal cells, preventing direct contact between the ENM and cell surface (Figure 3). Transparent, algae-shaped structures were also observed with few or no particles attached (Figure 4).

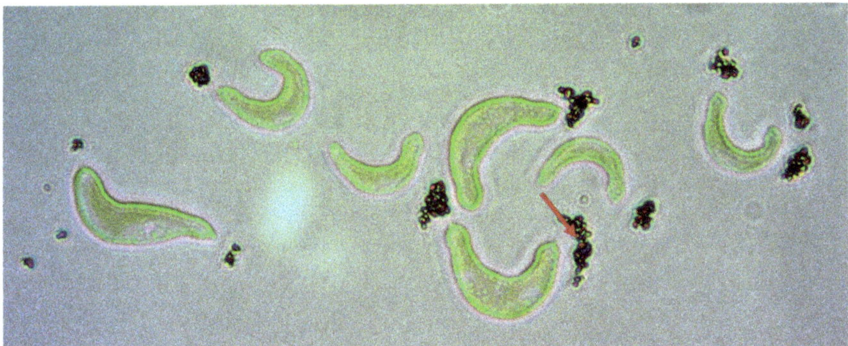

Figure 3. Nanoparticle attachment to transparent, sheath-like structures around algal cells (red arrow). This phase-contrast image (1000× magnification) was captured after 3 h incubation with 100 mg/L CeO_2 NM-213.

As described for the spike experiments, only a few particles attached to the algal cells during the growth inhibition tests after 72 h (Figures S5–S8). Accordingly, the shell-like structures around the algal cells were less obvious, although many transparent algae-shaped structures featuring a small number of attached particles were observed following the test period, especially at the higher test concentrations.

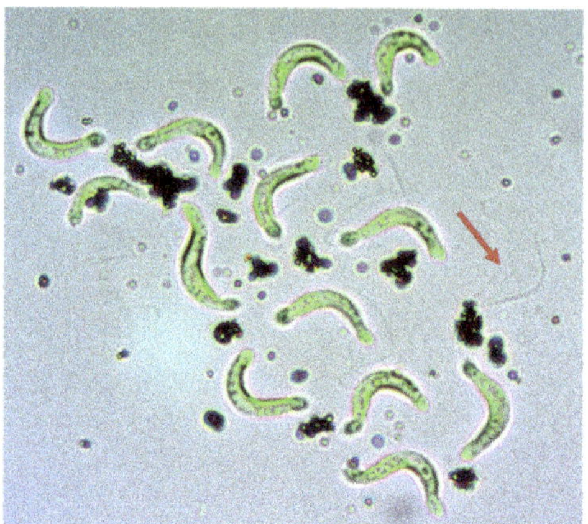

Figure 4. Transparent, algae-shaped structures (red arrow). This phase-contrast image (1000× magnification) was captured after 3 h incubation with 100 mg/L CeO_2 NM-213.

3.1.2. SEM-EDX Analysis

We studied the attachment of CeO_2 NM-212 to individual *R. subcapitata* cells by electron microscopy. Given the observed differences in toxicity (Section 3.3), we also compared particle attachment following incubation in two different media: GB and OECD. In both cases, we observed the attachment of CeO_2 NM-212 particles to algal cells and this was confirmed by EDX (Figure 5). Control cells that were not exposed to the ENM are shown in Figure S9. Interestingly, EDX analysis showed that not all particles associated with the algae were composed of cerium, but that iron and sodium were also present. The iron particles were deposited on the algae exclusively when the cells were incubated in GB medium and given that no iron-containing chemicals were used during sample preparation we conclude that iron in the GB medium precipitated onto the cells. Sodium precipitation was occasionally observed in the samples incubated in OECD medium. Furthermore, most cells were surrounded by filamentous web-like envelopes, probably the shrunken remains of extracellular polymeric structures (EPS) visible by light microscopy. Fewer of these structures were associated with the control cells (Figure S9). These network-like structures could also be artifacts generated during sample preparation.

The coverage of algal cells with NM-212 particles was quantified from the overlay of SEM and EDX images. This showed no significant difference between the GB and OECD media at an ENM concentration of 0.1 mg/L (2–3% coverage in each case), but a trend towards higher coverage in OECD medium (14% in OECD vs. 2.4% in GB) at an ENM concentration of 1 mg/L (Figure 6).

3.2. *Attachment of* TiO_2

The analysis of TiO_2 ENMs by light microscopy revealed that, like CeO_2 NM-211 and NM-212, the TiO_2 particles formed agglomerates that attached to the algal cells (Figures S10–S20). However, these particles formed heterogeneous agglomerates that differed in terms of compactness and the manner of attachment. Specifically, they primarily formed shell-like single layers of compact agglomerates around the algal cells but also formed loose agglomerations of cells and particles. Like the non-doped TiO_2 particles (Figure S10), the Eu-doped TiO_2 particles densely covered the algal cells and formed compact agglomerates up to 1 mm in diameter, as well as loose agglomerations with less-ordered attachments (Figures S11 and S12). In contrast, the Fe-doped TiO_2 particles formed only shell-like single layers of compact agglomerates (Figure S13). The gap between the TiO_2 NM-105 particles and algal cells was wider and the shell-like structure was less dense compared to the doped particles (Figure S14). NM-104 particles attached to the algae sparingly and formed a fragmented rather than a complete compact shell, and the width of the gap varied from indistinguishable up to a clearly defined space (Figure S15).

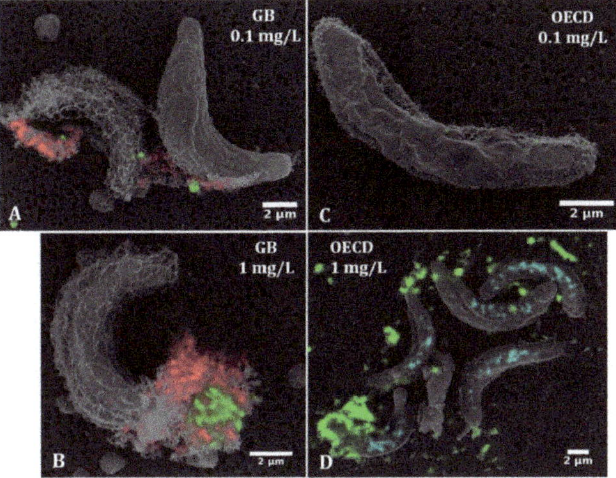

Figure 5. SEM-EDX (energy-dispersive X-ray spectroscopy) images of *Raphidocelis subcapitata* cells following exposure to CeO_2 NM-212 at two concentrations in two media, GB (**A**,**B**) and OECD (**C**,**D**). Some cells appear to be covered by network-like structures (potentially extracellular polymeric substances or artifacts that appear during sample preparation). Optically identical particulate assemblies are associated with the cells and, based on EDX analysis, contain not only cerium (green), but also sodium (blue) and iron (red).

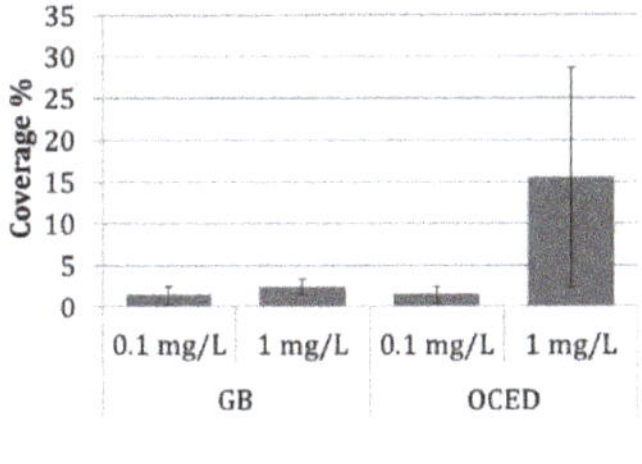

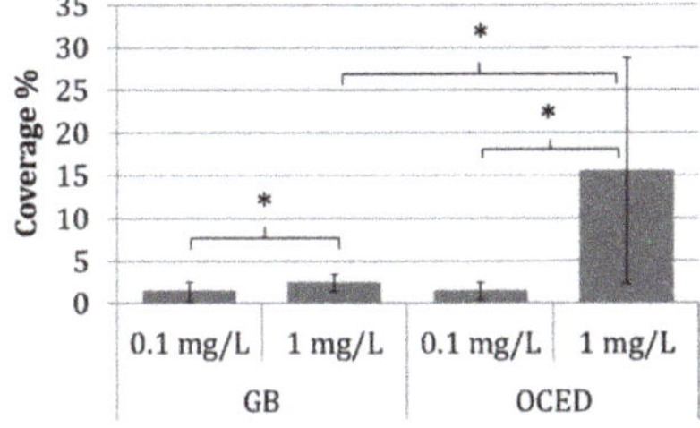

Figure 6. Coverage of the algal surface with cerium from NM-212 particles as determined by SEM-EDX image overlay analysis. We analyzed 5–6 samples per medium (GB and OECD), each sample comprised 2–7 cells. The data are means ± standard deviations. One-tailed unpaired *t*-test (Microsoft Excel 2013) revealed significant differences (marked by *) between the concentrations ($p = 0.033$ in GB medium, $p = 0.044$ in OECD medium) as well as between the media at 1 mg/L ($p = 0.045$).

Microscopic analysis during the growth inhibition tests revealed the formation of a shell-like layer for the highest test concentration of the Eu-doped TiO_2 particles (18 mg/L) and large but loose agglomerates at a lower concentration (2 mg/L) near the EC_{50} value (Figures S16 and S17). In contrast, the NM-104 agglomerates formed loose and open structures at the lowest test concentration of 7.5 mg/L (Figures S18–S20). Although most algal cells were incorporated into agglomerates, the status of individual cells was dependent on the ENM concentration, with few if any surface particles at the lowest test concentration (7.5 mg/L) or at 30 mg/L, but all algal cells incorporated into agglomerates at the highest test concentration of 120 mg/L.

3.3. Growth Inhibition Tests

For all three CeO_2 ENMs and two of the five TiO_2 ENMs, we carried out growth tests based on OECD test guideline 201 [12] and compared our results to earlier experiments [5]. We also compared the toxicity of CeO_2 NM-212, Eu-doped TiO_2, and non-doped TiO_2 in the two media and found that all three ENMs were more toxic in OECD medium than GB medium (Table 4).

Table 4. EC_{50} values determined in growth tests (OECD medium) with the green alga *Raphidocelis subcapitata* [1].

Nanomaterials	EC_{50} (mg/L)	EC_{50} (mg/L) Published Data [2]	EC_{50} (mg/L) [5] GB
CeO_2-NM-211	Not performed	8.5 [7.7–9.3]	Not performed
CeO_2 NM-212	10.9 [9.9–11.9] [4] 1.8 [n.d.] [5]	5.6 [3.0–10.4]	>100
CeO_2 NM-213	98.7 [96.5–101.4] [4]	43.8 [n.d.] [3]	Not performed
TiO_2 NM-104	126.9 [95.0 ± 190.4] [4]	62.6 [42.6–106]	Not performed
TiO_2 NM-105	Not performed	4.7 [3.5–5.5]	Not performed
TiO_2 Eu-doped	0.36 [0.34 ± 0.38] [4] 0.29 [n.d.] [5]	0.91 [0.75–1.10]	>100
TiO_2 Fe-doped	Not performed	3.6 [2.6–4.8]	Not performed
TiO_2 Non-doped	0.06 [n.d.] [5]	0.38 [0.33–0.43]	>100

[1] Values in brackets = confidence interval. [2] Tests carried out as described in Section 2 and results presented in the supporting information of a previous study [5]. [3] n.d. = not determined. [4] Data from this study (Fraunhofer Institute for Molecular Biology and Applied Ecology). [5] Data from this study (Helmholtz Centre for Environmental Research).

We compared the results generated by the two laboratories involved in this study (Fraunhofer Institute for Molecular Biology and Applied Ecology and Helmholtz Centre for Environmental Research) and also compared our results to published data [5]. In most cases, there was less than a five-fold variation. The ecotoxicity of the three CeO_2 ENMs differed by a factor of 100, with NM-213 showing the lowest toxicity. The ecotoxicity of the TiO_2 ENMs also differed by a factor of 100, with the non-doped and Eu-doped ENMs showing the highest toxicity and NM-104 the lowest.

3.4. Relationship between Attachment Behavior and Ecotoxicity

We observed a clear relationship between ecotoxicity and attachment efficiency for the three CeO_2 ENMs (Figure 7). NM-211 and NM-212 were highly toxic and also showed a great propensity for attachment to algal cells, whereas NM-213 was much less toxic and formed few agglomerations. In contrast, the relationship between ecotoxicity and attachment was more complex for the TiO_2 ENMs (Figure 7). There was no clear division between the particles that favored and disfavored interactions with algae, but rather a gradual change from strong to weak attachment. The Fe-doped ENM showed the strongest attachment, forming shell-like structures in compact agglomerates with most algal cells, followed by the Eu-doped and non-doped ENMs (mostly shell-like structures in compact agglomerates but some looser agglomerates), NM-105 (loose shell-like structures, gaps between algae and particles), and finally NM-104 (shell-like fragments, large gaps between algae and particles). The Eu-doped and non-doped ENMs showed the highest toxicity, followed by the Fe-doped ENM and NM-105, and finally NM-104. Furthermore, there was no obvious relationship between the crystalline structure of TiO_2 (rutile or rutile/anatase) and ecotoxicity.

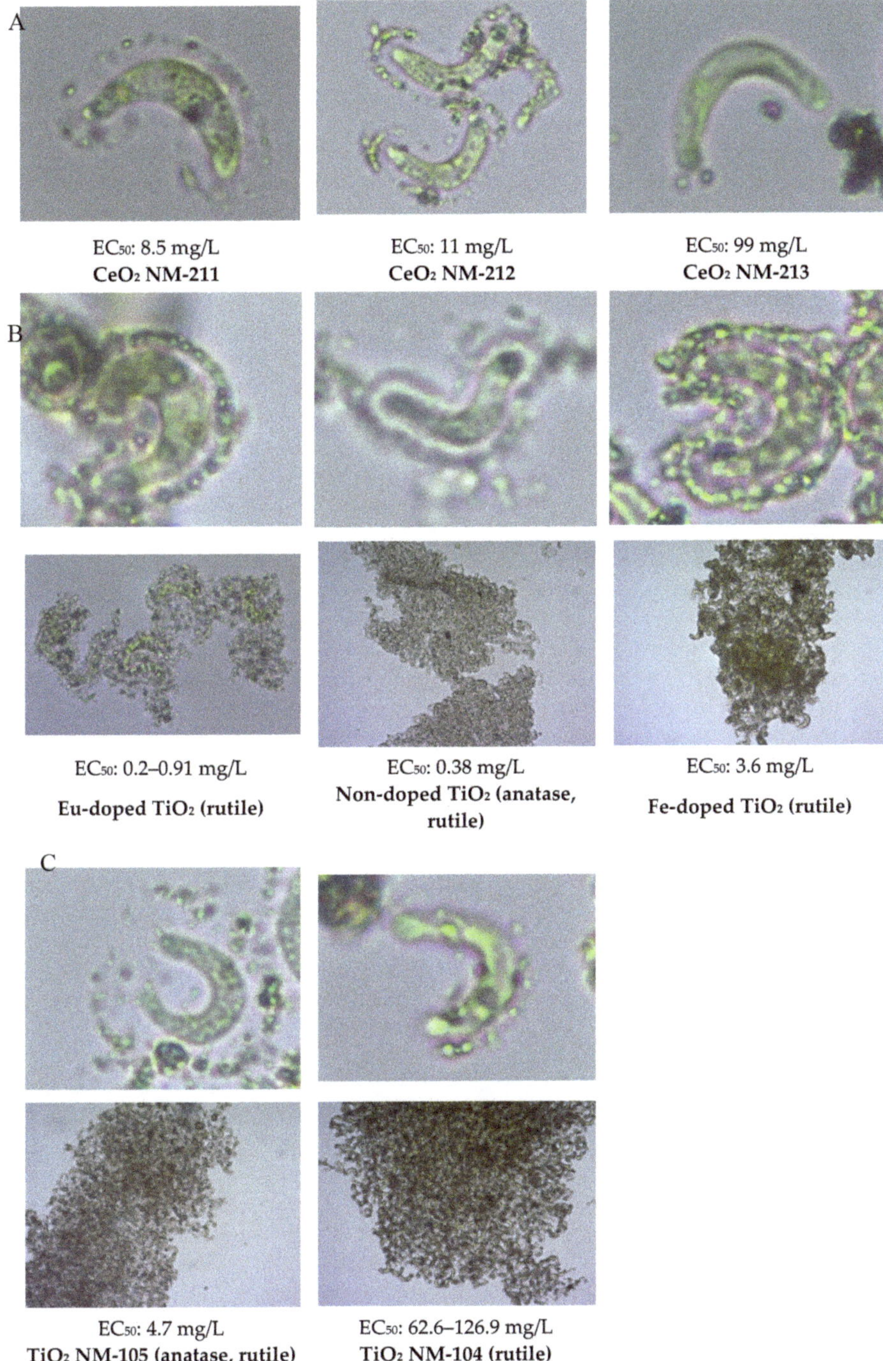

Figure 7. Comparative analysis of the EC_{50} values and attachment behavior of three CeO_2 ENMs (**A**) and five TiO_2 ENMs (**B**,**C**) (magnification 100× and 1000×).

4. Discussion

4.1. Attachment Behavior

Different attachment characteristics were observed depending on the type of ENM interacting with the algal cells. The ENMs in the spiking experiment did not attach directly to the algal surface but to a transparent, peripheral structure. Algae can excrete EPS [19] and increase the production of these molecules when under osmotic stress [20,21] or chemical stress caused by toxic chemicals [22–26] or natural toxin exudates [27]. Depending on their ability to bind the algal cell wall, EPS are generally categorized as soluble or bound substances [26,28]. They comprise many different organic acids, amino acids, peptides, sugars, polysaccharides, and oligosaccharides [20,28,29], and their composition varies among different species [30]. The role of EPS in the agglomeration of cells and ENMs has been reported before [31]. Their protective function has been demonstrated while investigating the toxicity of Ag nanoparticles toward the alga *Chlorella pyrenoidosa*, where EPS-extracted cells were more sensitive to the nanoparticles than control cells [29]. EPS production depends on the chemical substance that induces stress. For example, three ENMs (TiO_2, SiO_2, and CeO_2) were tested for their effect on the green alga *Dunaliella tertiolecta*, and although the EPS response in all cases was regulated by the Ca^{2+} signaling pathway, SiO_2 induced a 200–800% increase in EPS production whereas TiO_2 had only a limited effect [32]. A given ENM can also have different effects on EPS production in different algae: for example, antifouling agents induce a stronger EPS response in *Scenedesmus* sp. than *Chlorella* sp., mirroring the toxicity profiles against these species [33].

Existing test guidelines developed for chemical substances have been reevaluated to determine their suitability for ENMs, and any necessary adaptations have been identified. Furthermore, new test guidelines are under development and guidance documents have been provided [34]. One of the recommendations was a modified growth test with green algae [15,35]. We therefore considered modifications such as the use of fluorescence to determine algal cell numbers. The results of our growth tests using the same ENM under identical test conditions at different times and in different laboratories varied in most cases by less than a factor of five, confirming the robustness of the method and justifying its use to compare ecotoxicological data with attachment behavior (Table 4). Our results also varied (in most cases) by less than a factor of five compared to earlier experiments [5]. This is lower than the factor of 10 which is considered adequate for the definition of ecotoxicity categories for pesticides [36]. We also observed a clear relationship between the ecotoxicity of CeO_2 in the growth inhibition test and the attachment of ENMs to algae in the spiking experiment. The ENMs induced a shell-like structure in the short-term test, which was also observed in the growth test with NM-213. Such a structure was not observed in the control groups without ENMs, indicating that CeO_2 ENMs trigger two effects: first they induce the formation of a shell-like structure (presumably EPS) to protect the algae; and second, the nanoparticles adsorb to the EPS (depending on the type of ENM) which correlates with the toxicity in the growth test.

The interaction between ENMs and EPS is complex, probably reflecting the stressor-dependent composition of the EPS. For example, the infrared spectrum of the EPS induced by Ag differed from that induced by TiO_2 in the same algal species [33]. The adsorption of ENMs to algae is also influenced by surface modifications, as shown by the relatively stronger interactions between Ag nanoparticles and algae when the ENM was coated with polyvinylpyrrolidone rather than citrate [29]. Furthermore, P25 (14% rutile and 86% anatase) and anatase ENMs show higher affinity for proteins and polysaccharides whereas rutile ENMs attach more efficiently to phospholipids [37]. Anatase TiO_2 adsorbs more efficiently to bound EPS than rutile TiO_2, which may explain its greater ecotoxicity [38].

NM-105 has an anatase and rutile crystalline structure specifically designed for photocatalysis, and the remarkably wide gap between algae and the shell-like structure formed in the presence of NM-105 indicates a thick EPS layer, presumably induced by the increased production of reactive oxygen species. The combination of loose attachment (compared to the Fe-doped ENM) and the high reactivity (due to its photocatalytic activity) could explain the ecotoxicity of NM-105, which was comparable to the doped TiO_2 ENMs. The relationship between EPS induction and the toxicity of TiO_2 and Ag was reported previously [33] and the underlying causes of the interaction between EPS and metal(loid)s have been comprehensively reviewed [28].

The evidence from our study and previous reports therefore suggests that attachment behavior can be used as a surrogate parameter to indicate ecotoxicity. The attachment of ENMs to algal cells is merely an observation of interaction behavior, and although our ENMs were characterized in detail, the physicochemical properties responsible for the different attachment behaviors could not be identified. There is no obvious direct relationship between attachment behavior and parameters such as primary particle size, surface area, morphology, surface chemistry, crystalline structure, zeta-potential, isoelectric point, agglomeration size, reactivity, solubility or ecotoxicity [4,5].

The growth of the algae in our experiments was inhibited by the ENMs, indicating a toxic effect despite the production of protective extracellular substances [28]. The toxicity of Ag ENMs may reflect the penetration of the cell wall and plasma membrane followed by the intracellular release of Ag^+ [29]. Similarly, TiO_2 ENMs may penetrate the cell and inhibit the synthesis of chlorophyll *b* and carotenoids [9]. However, TiO_2 ENMs also promote significant hydrophobic interactions by forming complexes of titanium acetate between the aliphatic –COOH groups of EPS and titanium ions [38]. This involves the significant accumulation of TiO_2 ENMs on the EPS but almost no penetration, yet toxic effects were nevertheless observed (particularly for the anatase ENM compared to the rutile ENM, corresponding to the extent of adsorption) [38]. The mode of action is unclear but may involve the reduced penetration of light caused by the shell of adsorbed nanoparticles. Shading due to sorption is considered a nanospecific effect in contrast to shading by turbid test dispersions [35]. Toxicity could also reflect the increased energy demand due to the production of EPS, thus reducing the growth rate.

A spike experiment addressing attachment and agglomeration behavior can only be considered as a proxy for ecotoxicity. We found that the agglomeration behavior depends on the concentration ratio between the algal cells and ENM, but in the spike experiment the concentration of both components was high. Furthermore, the EC_{50} values of the ENM vary by several orders of magnitude, indicating that the concentration ratio changes during the 72-h growth test as the algal cell number increases while the ENM concentration stays the same. The high concentration of the algae and ENMs in the spike test therefore differs from the conditions in the growth inhibition test. The stronger induction of EPS production may explain the different behavior of the toxic NM-212 particles in the growth inhibition and spike tests, whereas the discrepancy was minimal for the less toxic NM-213 particles. However, the relationship between attachment and ecotoxicity is plausible, and spike experiments, despite their weaknesses, could therefore be useful to determine the ecotoxicity of ENMs toward algae. The ENM characterization and screening experiments revealed similar sizes for the agglomerates. For example, the agglomerate size for NM-212 was 830 ± 200 nm in the characterization experiments (Table 1) and ~1 μm in the screening assay (Figure 1). We used the same ENM concentration in both cases (100 mg/L in OECD medium). This confirms that no significant artifacts were introduced during the preparation of samples.

For NM-212, the quantification of surface coverage on the algal cells indicated a strong correlation between attachment and toxicity. In GB medium, the EC_{50} for the CeO_2 particles was >100 mg/L and the coverage was ~2%, whereas in OECD medium the EC_{50} for the same material was 1.8 mg/L and the coverage was ~14%. The difference in toxicity may reflect the compositions of the two media, with the rich GB medium containing more components at higher concentrations hence providing better growth conditions (Table 2). The coverage is likely to be underestimated (while preserving the trend) because particle detachment may occur during sample preparation. The light microscopy data indicate higher coverage due to the higher concentrations of ENMs used in the spike assays (100 mg/L compared to 1 mg/L). However, the detection of sodium and iron precipitates on the cell surface indicates that particles remain attached during fixation. Sodium precipitates were detected on cells after incubation in OECD medium, which is rich in sodium (Table 2), whereas iron was detected in cells incubated in GB medium, which contains Fe-EDTA. The presence of medium-specific elements on the cell surface after fixation indicates that particle attachment is strong enough to endure multiple washing steps, although it is possible that fixation may have a differential effect on the elements iron, sodium, and cerium.

The electron microscopy data were inconclusive regarding the modulation of EPS formation by different ENMs in the two media. A network-like structure was clearly present on the surface of the algal cells, but there was no clear difference between cells exposed to different ENMs, between ENM-exposed cells and controls, or between cells incubated in GB and OCED media. The fixation procedure may be too harsh to preserve EPS structures in a state that allows their thickness to be compared across different treatment groups.

4.2. Read-Across and Grouping Strategies

In the read-across technique, endpoint information for one chemical is used to predict the same endpoint for another, which is considered similar in some way [39]. For nanomaterials, the approach is only permitted in the context of various forms of the same substance [1]. Additionally, grouping is an essential part of the ECHA draft guidance on the registration of sets of similar nanoforms: "... A 'set of similar nanoforms' is a group of nanoforms ... where the clearly defined boundaries in the parameters ... of the individual nanoforms within the set still allow to conclude that the hazard assessment ... of these nanoforms can be performed jointly. A justification shall be provided to demonstrate that a variation within these boundaries does not affect the hazard assessment ... of the similar nanoforms in the set" [40]. In the grouping and read-across concept, a nanoform is an ENM with the same crystalline structure, comparable particle size distribution, morphology, surface functionalization, and surface area [40]. Sufficient similarity can be assumed if there is preferably a logical ranking of the materials that allows the identification of a worst case [40].

Given the different attachment and agglomeration behaviors of the TiO_2 and CeO_2 ENMs, read-across may solely be possible between various forms of the same ENM. The ECHA does not define threshold values for physicochemical parameters, and the three CeO_2 nanomaterials could therefore be assigned to two groups. However, different groupings arise from the selection of different physicochemical parameters: based on the primary particle size, NM-211 is smaller than NM-212 and NM-213; based on the surface area, NM-213 differs from NM-211 and NM-212; and in the algal ecotoxicity tests, NM-211 and NM-212 are much more toxic than NM-213. Therefore, grouping based on individual physicochemical parameters is considered to be less targeted. We still lack parameters with a clear relationship to the test organisms, but in algae the attachment behavior of the ENMs may provide an acceptable proxy. For the three CeO_2 ENMs, there was a clear correlation between attachment behavior and ecotoxicity.

The selected physicochemical properties of TiO_2 ENMs also affect the results of grouping. The five ENMs form one group based on size, but grouping by surface area leads to the separation of Eu-doped TiO_2, grouping by reactivity leads to the separation of NM-105, and grouping by crystalline structure separates the rutile/anatase NM-105 and Fe-doped TiO_2 from the three rutile ENMs. NM-104 can also form a separate group as the only coated ENM. If all these differences are considered at the same time, the five TiO_2 ENMs serve as representatives of five groups. However, this is not justified for the assessment of ecotoxicity toward algae. Based on the empirical data, we have three groups differing in ecotoxicity by a factor of 10, but they have heterogeneous physicochemical properties. We observed comparable ecotoxicity for ENMs differing in crystalline structure and reactivity (NM-105 is anatase/rutile and reactive, whereas Fe-doped TiO_2 is rutile and non-reactive) or differing in surface area (Eu-doped TiO_2 = 148 m^2/g, whereas non-doped TiO_2 = 78 m^2/g). As discussed for CeO_2, attachment behavior appears to be a better indicator of similar ecotoxicity toward algae than physicochemical parameters. The materials with the strongest attachment behavior and lowest reactivity (Eu-doped and non-doped TiO_2) showed the greatest ecotoxicity, followed by the material with the highest reactivity and moderate attachment efficiency (NM-105). The large gap between the ENMs and algal cells, indicating a shell-like structure for protection, can be explained by the high reactivity of the ENM. The material with low attachment efficiency and low reactivity (NM-104) showed the lowest ecotoxicity. Only the ecotoxicity of Fe-doped TiO_2 cannot be explained by the attachment behavior and reactivity. A higher ecotoxicity than Eu-doped and non-doped TiO_2 would be expected, and additional work is required to improve the prediction of the ecotoxicity for this material and corresponding ENMs. The underlying properties responsible for the different attachment behaviors are still unclear, but until these are identified it should be possible to use the easily-measured parameter of attachment behavior as a proxy for ecotoxicity, at least in the case of CeO_2 and TiO_2 ENMs.

5. Conclusions

In this study, we used light microscopy and SEM-EDX to investigate the attachment of three CeO_2 and five TiO_2 ENMs to the green alga *R. subcapitata* and compared the attachment behavior with ecotoxicity data based on a growth inhibition assay following OECD test guideline 201. Light microscopy allowed us to screen the algae without compromising the structural integrity of the sample, whereas SEM-EDX provided images of greater resolution combined with elemental analysis. We found that CeO_2 and TiO_2 ENMs induce the formation of EPS by the algae and we observed a relationship between ecotoxicity (based on growth inhibition data) and the attachment behavior of the ENMs. In contrast, there was no simple relationship between algal ecotoxicity and the physicochemical properties of the ENMs. Attachment behavior can therefore be considered as a valuable proxy for the identification of sets of similar nanoforms in the context of grouping and read-across, which reduce the number of tests required for risk assessment. Our observations are based on the analysis of sparingly soluble, spherical ENMs available as white powder. Further experiments are required to determine whether attachment behavior has a similar predictive power for the ecotoxicity of different ENM shapes (such as rods, fibers, or platelets) and colors (such as red Fe_2O_3). Colored materials indicate the selective reflection of different wavelengths of light, so we cannot exclude potential interference with the pigments involved in photosynthesis and algal growth. Furthermore, ENMs that release ions such as Ag^+ should also be tested.

Supplementary Materials: The following are available online at http://www.mdpi.com/2079-4991/10/6/1021/s1.

Author Contributions: Conceptualization, K.H.-R., M.S., and D.K.; funding acquisition, K.H.-R. and D.K.; investigation, T.S., C.N., and H.P.D.; writing—original draft preparation, K.H.-R., K.S., C.N., M.S., and D.K.; writing—review and editing, K.H.-R., K.S., C.N., H.P.D., M.S., and D.K. All authors have read and agreed to the published version of the manuscript.

Funding: The experimental work was funded by the German Federal Ministry of Education and Research (BMBF) under the joint project nanoGRAVUR (grant no. 03XP0002). For the modification of the grouping approach by the integration of attachment efficiency data, this project received funding from the European Union Horizon 2020 research and innovation programme under grant agreement no. 814530 (NANORIGO).

Acknowledgments: The authors are grateful for access to the scanning electron microscope at ProVIS – Centre for Chemical Microscopy at the Helmholtz Centre for Environmental Research, Leipzig, which is supported by European Regional Development Funds (EFRE–Europe funds Saxony) and the Helmholtz Association. Technical assistance with the green algae culture and toxicity testing by Silke Aulhorn (Helmholtz Centre for Environmental Research) is gratefully acknowledged. The authors would like to thank Richard M. Twyman for reviewing the manuscript.

Conflicts of Interest: The authors declare no conflict of interest. The funders had no role in the design of the study, in the collection, analysis, or interpretation of data, in the writing of the manuscript, or in the decision to publish the results.

References

1. ECHA. *Guidance on Information Requirements and Chemical Safety Assessment—Appendix R.6.1 for Nanomaterials Applicable to the Guidance on QSARs and Grouping of Chemicals*; ECHA: Helsinki, Finland, 2017.
2. Arts, J.H.E.; Irfan, M.-A.; Keene, A.M.; Kreiling, R.; Lyon, D.; Maier, M.; Michel, K.; Neubauer, N.; Petry, T.; Sauer, U.G.; et al. Case studies putting the decision-making framework for the grouping and testing of nanomaterials (DF4nanoGrouping) into practice. *Regul. Toxicol. Pharmacol.* **2016**, *76*, 234–261. [CrossRef] [PubMed]

3. Oomen, G.A.; Bleeker, A.E.; Bos, M.P.; Van Broekhuizen, F.; Gottardo, S.; Groenewold, M.; Hristozov, D.; Hund-Rinke, K.; Irfan, M.-A.; Marcomini, A.; et al. Grouping and Read-Across Approaches for Risk Assessment of Nanomaterials. *Int. J. Environ. Res. Public Health* **2015**, *12*, 13415–13434. [CrossRef] [PubMed]
4. Hund-Rinke, K.; Schlich, K.; Kühnel, D.; Hellack, B.; Kaminski, H.; Nickel, C. Grouping concept for metal and metal oxide nanomaterials with regard to their ecotoxicological effects on algae, daphnids and fish embryos. *NanoImpact* **2018**, *9*, 52–60. [CrossRef]
5. Kuehnel, D.; Nickel, C.; Zalm, E.V.D.; Kussatz, C.; Herrchen, M.; Meisterjahn, B.; Hund-Rinke, K. Closing gaps for environmental risk screening of engineered nanomaterials. *NanoIMPACT* **2019**, *15*, 100173. [CrossRef]
6. Metzler, D.M.; Li, M.H.; Erdem, A.; Huang, C.P. Responses of algae to photocatalytic nano-TiO_2 particles with an emphasis on the effect of particle size. *Chem. Eng. J.* **2011**, *170*, 538–546. [CrossRef]
7. Aruoja, V.; Dubourguier, H.C.; Kasemets, K.; Kahru, A. Toxicity of nanoparticles of CuO, ZnO and TiO_2 to microalgae Pseudokirchneriella subcapitata. *Sci. Total Environ.* **2009**, *407*, 1461–1468. [CrossRef]
8. Ozkaleli, M.; Erdem, A. Biotoxicity of TiO_2 Nanoparticles on Raphidocelis subcapitata Microalgae Exemplified by Membrane Deformation. *Int. J. Environ. Res. Public Health* **2018**, *15*, 416. [CrossRef] [PubMed]
9. Chen, L.; Zhou, L.; Liu, Y.; Deng, S.; Wu, H.; Wang, G. Toxicological effects of nanometer titanium dioxide (nano-TiO_2) on Chlamydomonas reinhardtii. *Ecotoxicol. Environ. Saf.* **2012**, *84*, 155–162. [CrossRef]
10. Rasmussen, K.; Mast, J.; De Temmerman, P.-J.; Verleysen, E.; Waegeneers, N.; van Steen, F.; Pizzolon, J.; de Temmerman, L.; Van Doren, E.; Jensen, K.A.; et al. *Titanium Dioxide, NM-100, NM-101, NM-102, NM-103, NM-104, NM-105: Characterisation and Physico-Chemical Properties*; European Commission, Joint Research Centre, Insitute for Health and Consumer Production: Ispra, Italy, 2014.
11. Singh, C.; Friedrichs, S.; Ceccone, G.; Gibson, N.; Jensen, K.A.; Levin, M.; Infante, H.G.; Carlender, D.; Rasmussen, K. *Cerium Dioxide, NM-211, NM-212, NM-213. Characterisation and Test item Preparation. JRC Repository: NM-Series of Representative Manufactured Nanomaterials*; Publications Office of the European Union: Ispra, Italy, 2014.
12. OECD Guideline 201. *OECD Guidelines for the Testing of Chemicals. Test Guideline 201: Freshwater Alga and Cyanobacterial, Growth Inhibition Test*; Organisation for Economic Co-operation and Development: Paris, France, 2011; Available online: http://www.oecd-ilibrary.org/docserver/download/9720101e.pdf?expires=1497009075&id=id&accname=guest&checksum=B05A403DAB471E46DEDD582972FEDABE (accessed on 20 March 2018).
13. Shi, T.; Schins, R.P.F.; Knaapen, A.M.; Kuhlbusch, T.; Pitz, M.; Heinrich, J.; Borm, P.J. Hydroxyl radical generation by electron paramagnetic resonance as a new method to monitor ambient particulate matter composition. *J. Environ. Monit.* **2003**, *5*, 550–556. [CrossRef]
14. Lipovsky, A.; Tzitrinovich, Z.; Friedmann, H.; Applerot, G.; Gedanken, A.; Lubart, R. EPR study of visible light-induced ROS generation by nanoparticles of ZnO. *J. Phys. Chem. C* **2009**, *113*, 15997–16001. [CrossRef]
15. Hund-Rinke, K.; Baun, A.; Cupi, D.; Fernandes, T.F.; Handy, R.; Kinross, J.H.; Navas, J.M.; Peijnenburg, W.; Schlich, K.; Shaw, B.J.; et al. Regulatory ecotoxicity testing of nanomaterials—Proposed modifications of OECD test guidelines based on laboratory experience with silver and titanium dioxide nanoparticles. *Nanotoxicology* **2016**, *10*, 1442–1447. [CrossRef] [PubMed]
16. Faust, M.R.; Altenburger, R.; Bödeker, W.; Grimme, L. Algentoxizitätstest mit synchronisierten kulturen. *Schriftenreihe Ver. Wasser Boden Lufthyg.* **1992**, *89*, 311–329.
17. Rohde, F. Entwicklung eines Nichtlinearen Algorithmus zur Automatischen Elastischen Registrierung Kor-Relativer Mikroskopiedaten. Master's Thesis, Leipzig University of Applied Sciences, Leipzig, Germany, 2018.
18. Rohde, F.; Braumann, U.-D.; Schmidt, M. Correlia: An ImageJ plug-in to register and visualise multi-modal correlative micrographs. in preparation. In Proceedings of the Berlin Microscope Conference, Berlin, Germany, 1–5 September 2019.
19. Xiao, R.; Zheng, Y. Overview of microalgal extracellular polymeric substances (EPS) and their applications. *Biotechnol. Adv.* **2016**, *34*, 1225–1244. [CrossRef] [PubMed]
20. Maršálek, B.; Rojíčková, R. Stress Factors Enhancing Production of Algal Exudates: A Potential Self-Protective Mechanism? *Z. Naturforsch* **1996**, *51*, 646–650. [CrossRef]
21. Maršálek, B.; Zahradníčková, H.; Hronková, M. Extracellular Production of Abscisic Acid by Soil Algae under Salt, Acid or Drought Stress. *Z. Naturforsch* **1992**, *47*, 701–704. [CrossRef]

22. Iswarya, V.; Sharma, V.; Chandrasekaran, N.; Mukherjee, A. Impact of tetracycline on the toxic effects of titanium dioxide (TiO_2) nanoparticles towards the freshwater algal species, Scenedesmus obliquus. *Aquat. Toxicol.* **2017**, *193*, 168–177. [CrossRef]
23. Miao, A.J.; Schwehr, K.A.; Xu, C.; Zhang, S.J.; Luo, Z.P.; Quigg, A.; Santschi, P.H. The algal toxicity of silver engineered nanoparticles and detoxification by exopolymeric substances. *Environ. Pollut.* **2009**, *157*, 3034–3041. [CrossRef]
24. Morelli, E.; Gabellieri, E.; Bonomini, A.; Tognotti, D.; Grassi, G.; Corsi, I. TiO_2 nanoparticles in seawater: Aggregation and interactions with the green alga Dunaliella tertiolecta. *Ecotoxicol. Environ. Saf.* **2018**, *148*, 184–193. [CrossRef]
25. Wang, C.Q.; Dong, D.M.; Zhang, L.W.; Song, Z.W.; Hua, X.Y.; Guo, Z.Y. Response of Freshwater Biofilms to Antibiotic Florfenicol and Ofloxacin Stress: Role of Extracellular Polymeric Substances. *Int. J. Environ. Res. Public Health* **2019**, *16*, 715. [CrossRef]
26. Zhao, J.F.; Liu, S.X.; Liu, N.; Zhang, H.; Zhou, Q.Z.; Ge, F. Accelerated productions and physicochemical characterizations of different extracellular polymeric substances from Chiorella vulgaris with nano-ZnO. *Sci. Total Environ.* **2019**, *658*, 582–589. [CrossRef]
27. El-Sheekh, M.M.; Khairy, H.M.; El-Shenody, R. Algal production of extra and intra-cellular polysaccharides as an adaptive response to the toxin crude extract of Microcystis aeruginosa. *Iran. J. Environ. Health Sci. Eng.* **2012**, *9*, 10. [CrossRef]
28. Naveed, S.; Li, C.; Lu, X.; Chen, S.; Yin, B.; Zhang, C.; Ge, Y. Microalgal extracellular polymeric substances and their interactions with metal(loid)s: A review. *Crit. Rev. Environ. Sci. Technol.* **2019**, *49*, 1769–1802. [CrossRef]
29. Zhou, K.; Hu, Y.; Zhang, L.; Yang, K.; Lin, D. The role of exopolymeric substances in the bioaccumulation and toxicity of Ag nanoparticles to algae. *Sci. Rep.* **2016**, *6*, 32998. [CrossRef] [PubMed]
30. Taylor, C.; Matzke, M.; Kroll, A.; Read, D.S.; Svendsen, C.; Crossley, A. Toxic interactions of different silver forms with freshwater green algae and cyanobacteria and their effects on mechanistic endpoints and the production of extracellular polymeric substances. *Environ. Sci.-Nano* **2016**, *3*, 396–408. [CrossRef]
31. Yang, Y.; Hou, J.; Wang, P.; Wang, C.; Wang, X.; You, G. Influence of extracellular polymeric substances on cell-NPs heteroaggregation process and toxicity of cerium dioxide NPs to Microcystis aeruginosa. *Environ. Pollut.* **2018**, *242*, 1206–1216. [CrossRef]
32. Chiu, M.H.; Khan, Z.A.; Garcia, S.G.; Le, A.D.; Kagiri, A.; Ramos, J.; Tsai, S.M.; Drobenaire, H.W.; Santschi, P.H.; Quigg, A.; et al. Effect of Engineered Nanoparticles on Exopolymeric Substances Release from Marine Phytoplankton. *Nanoscale Res. Lett.* **2017**, *12*. [CrossRef]
33. Natarajan, S.; Lakshmi, D.S.; Bhuvaneshwari, M.; Iswarya, V.; Mrudula, P.; Chandrasekaran, N.; Mukherjee, A. Antifouling activities of pristine and nanocomposite chitosan/TiO_2/Ag films against freshwater algae. *RSC Adv.* **2017**, *7*, 27645–27655. [CrossRef]
34. Rasmussen, K.; Rauscher, H.; Kearns, P.; González, M.; Riego Sintes, J. Developing OECD test guidelines for regulatory testing of nanomaterials to ensure mutual acceptance of test data. *Regul. Toxicol. Pharmacol.* **2019**, *104*, 74–83. [CrossRef]
35. OECD. *Draft-Guidance Document on Aquatic and Sediment Toxicological Testing of Nanomaterials*; OECD: Paris, France, 2019.
36. Environmental Protection Agency of United States. Technical Overview of Ecological Risk Assessment—Analysis Phase: Ecological Effects Characterization. Available online: https://www.epa.gov/pesticide-science-and-assessing-pesticide-risks/technical-overview-ecological-risk-assessment-0#Ecotox (accessed on 12 December 2019).
37. Li, K.; Qian, J.; Wang, P.; Wang, C.; Fan, X.; Lu, B.; Tian, X.; Jin, W.; He, X.; Guo, W. Toxicity of Three Crystalline TiO_2 Nanoparticles in Activated Sludge: Bacterial Cell Death Modes Differentially Weaken Sludge Dewaterability. *Environ. Sci. Technol.* **2019**, *53*, 4542–4555. [CrossRef] [PubMed]
38. Gao, X.; Zhou, K.; Zhang, L.; Yang, K.; Lin, D. Distinct effects of soluble and bound exopolymeric substances on algal bioaccumulation and toxicity of anatase and rutile TiO_2 nanoparticles. *Environ. Sci. Nano* **2018**, *5*, 720–729. [CrossRef]

39. ECHA. *Guidance on Information Requirements and Chemical Safety Assessment Chapter R.6: QSARs and Grouping of Chemicals*; ECHA: Helsinki, Finland, 2008; Available online: https://echa.europa.eu/documents/10162/13632/information_requirements_r6_en.pdf/77f49f81-b76d-40ab-8513-4f3a533b6ac9 (accessed on 7 June 2019).
40. ECHA. *Appendix for Nanoforms Applicable to the Guicance on Registration and Substance Identification*; Draft (Public) Version 1.0.; ECHA: Helsinki, Finland, 2019.

 *nanomaterials*

Review

Effects of Airborne Nanoparticles on the Nervous System: Amyloid Protein Aggregation, Neurodegeneration and Neurodegenerative Diseases

Anna von Mikecz * and Tamara Schikowski

IUF—Leibniz Research Institute for Environmental Medicine gGmbH, Heinrich-Heine-University, 40225 Duesseldorf, Germany; Tamara.Schikowski@IUF-Duesseldorf.de

* Correspondence: mikecz@uni-duesseldorf.de; Tel.: +49-(0)211-3389-358

Received: 7 June 2020; Accepted: 7 July 2020; Published: 10 July 2020

Abstract: How the environment contributes to neurodegenerative diseases such as Alzheimer's is not well understood. In recent years, science has found augmenting evidence that nano-sized particles generated by transport (e.g., fuel combustion, tire wear and brake wear) may promote Alzheimer's disease (AD). Individuals residing close to busy roads are at higher risk of developing AD, and nanomaterials that are specifically generated by traffic-related processes have been detected in human brains. Since AD represents a neurodegenerative disease characterized by amyloid protein aggregation, this review summarizes our current knowledge on the amyloid-generating propensity of traffic-related nanomaterials. Certain nanoparticles induce the amyloid aggregation of otherwise soluble proteins in in vitro laboratory settings, cultured neuronal cells and vertebrate or invertebrate animal models. We discuss the challenges for future studies, namely, strategies to connect the wet laboratory with the epidemiological data in order to elucidate the molecular bio-interactions of airborne nanomaterials and their effects on human health.

Keywords: air pollution; Alzheimer's disease; amyloid; *Caenorhabditis elegans*; COVID-19; dementia; neurotoxicology; particulate matter; serotonin; tire wear

1. Introduction

Airborne particles constitute a threat to human health. Especially, the nanosized particle fraction that is highly abundant in the urban atmosphere has the ability to penetrate virtually all organs and possesses high bioreactivity. They have also been linked to respiratory viral infections such as the SARS-CoV-2 virus and influenza as well as other respiratory and cardiovascular diseases. Recent work links combustion- and friction-derived air pollution nanoparticles (airNPs) not only to adverse health effects of the respiratory and cardiovascular systems, but also to neurodevelopmental and cognitive impairment. Consistent with this idea, this review focuses on the aging nervous system as a target of airNPs. This particularly includes novel findings concerning neurodegenerative bio-interactions of traffic-related nanomaterials in the invertebrate animal model *Caenorhabditis elegans*, which possesses a simple yet highly informative nervous system and can be investigated over its entire lifespan (i.e., enables the whole-life investigation of chronic exposures to pollutants). Medium-throughput analyses of the roundworm *C. elegans* cultivated in 96-well microtiter plate formats allow the exploration of diverse nanoparticles and their properties, including those pre- and post-use. Direct collaborations between model organism researchers and epidemiologists are suggested to identify cellular pathways of neurotoxic airNPs and thereby promote the neurosafety of nanomaterials.

1.1. Potential Association of Combustion- and Friction-Derived airNPs with Neurodegenerative Aggregation Diseases such as Alzheimer's Disease

The exact mechanisms of neuronal death in neurodegenerative diseases such as Alzheimer's disease (AD) and Parkinson's disease (PD) are largely unknown. Studies on air pollution exposure with cardiovascular and cerebrovascular diseases suggest a harmful impact on the brain and cognitive processes through vascular and inflammatory mechanisms [1] However, the extent to which air pollution can affect cognitive decline and dementia in the elderly is not fully understood. This is despite the fact that AD as well as PD represent a growing health problem in the aging population globally.

From the set of existing explanatory models, there is compelling genetic evidence for the aging and functional loss of protein homeostasis in cells of the central nervous system (CNS) that contributes to degenerative phenotypes. A disturbed balance between protein synthesis, folding, and degradation induces the abnormal protein aggregation in neural cells that can go as far as the formation of toxic oligomers and amyloid protein structures [2,3]. These amyloid structures are characterized by insolubility that above a certain threshold is refractory to the cellular protein degradation pathways. Amyloid protein aggregation represents a common feature of the neuropathology in AD and PD, and is closely associated with the expression of amyloid-β peptide, tau protein and α-synuclein, respectively.

In addition to aging as a risk factor for the induction of AD and PD, the contribution of environmental factors such as certain pollutants is given consideration. While case and epidemiologic studies link the premature onset of PD with pesticides or cohorts of occupationally exposed welders [4], AD has recently been correlated with urban air pollution, specifically particulate matter (PM) [5–7]. A meta-analysis of four significant cohorts in Great Britain, Canada, the USA and Taiwan revealed a positive association between the exposure to air pollution PM and dementia (e.g., AD). The Canadian study showed a positive association between a person's domicile located within 50–300 m of a busy road and newly diagnosed cases of dementia with a hazard ratio of 1.12 and a 95% confidence interval of 1.10–1.14 [8]. Notably, previous studies identified the key exposure zone of traffic-related nanoparticles within 500 m and critically within 50 m from the traffic route [9,10]. The inhalation of air pollution and diesel exhaust was shown to induce inflammatory changes as well as hallmarks of AD, including amyloid formation [11–13] (*1.2*). A recent review summarized the results from epidemiological studies indicating that exposure to air pollution can have adverse effects on cognitive decline and impairment [14].

A new and emerging angle of urban air pollution and its adverse health effects is the contribution of rising temperatures due to climate change, especially in cities. Metropolitan areas represent vulnerable targets of the climate crisis since their buildings and pavements absorb sunlight and raise local temperatures, which in turn promote the phenomenon of urban heat islands [15]. Additionally, climate change promotes an urban microclimate that is characterized by the increase of extreme events such as the number and duration of heat waves [16]. As we know little about the interactions between a heated urban microclimate and the adverse health effects of traffic-related nanoparticulate air pollution, respective analyses assessing the role of temperature in the promotion of adverse health effects of airNPs such as neurodegeneration, that is, neurodegenerative diseases, are much needed.

1.2. Entry Portals—Where airNPs Have Been Found

Consistent with the idea of the urban atmosphere as a risk factor for dementia and AD, post-mortem brain samples from clinically healthy humans and dogs exposed to lifetime air pollution while living in the metropolitan areas of Mexico City or Manchester (UK) displayed typical hallmarks of AD pathogenesis, that is, aberrant deposition of amyloid-β peptide and tau protein [17]. Moreover, electron microscopy and magnetic analyses identified the presence of metal-bearing NPs, including mixed Fe^{2+}/Fe^{3+} (magnetite), that represent specific combustion emissions. Of particular concern is the association of air pollution, combustion- and friction-derived NPs in young populations (i.e., children and young adults living in major cities, [6]). Rodent animal models of urban nanoparticulate air pollution show the consistent induction of inflammatory responses in major brain regions, increased DNA damage in cell nuclei of central neurons, and increased levels of AD-related tau phosphorylation [11,13].

1.3. Combustion- and Friction-Derived airNPs

Traffic-derived emissions are a major source of urban PM, constituting up to 80% of airborne concentrations of PM in the urban environment ([18]; Figure 1). A recent review by Gonet and Maher gives a comprehensive summary of the generation, composition and environmental distribution of transport-related particle pollution [17]. The paper likewise displays transmission electron microscopy micrographs of the most prevalent nanomaterials that are generated by fuel combustion or tire and brake wear. Air pollution nanoparticles originate from exhaust emissions such as diesel, gasoline and kerosene, but likewise from brake and tire wear [17]. Certain airNPs such as nano-sized silicon dioxide particles (nano silica) constitute both brake and tire wear [17]. Car tires profit from the addition of silica nanomaterials with regard to enhancement and durability. In contrast, NPs such as nano ceria specifically surface in diesel vehicle exhaust, due to their application as fuel additives ([19]; Figure 1, inset). In bench tests, the addition of nano ceria reduced diesel exhaust emissions of CO_2, CO and total particulate mass in a ceria-concentration-dependent manner; however, emissions of other pollutants such as NOx (+9.3%) and the fraction of highly bioactive nanoparticulate particles (+32%) were simultaneously increased. This clearly shows that airNPs may not only pose a health hazard as a result of their intrinsic bio-interactions, but likewise via increasing the bioavailable concentrations of other traffic-related pollutants during the combustion process.

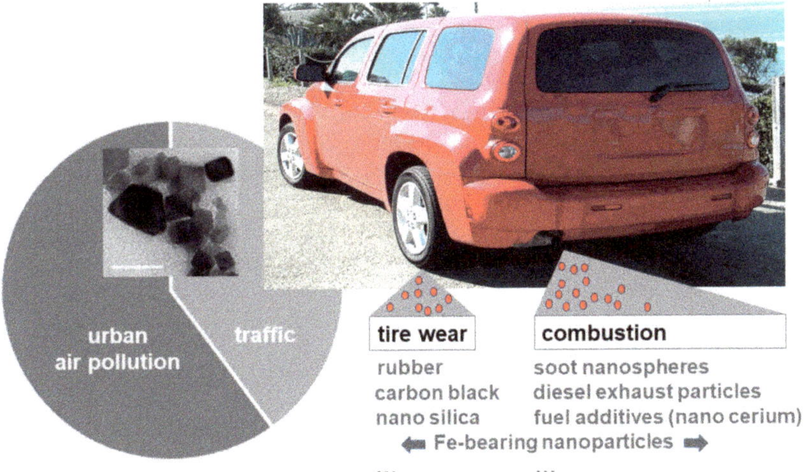

Figure 1. Schematic of particulate matter (PM) emissions. Traffic-related particle emissions constitute a significant proportion of urban air pollution (pie chart). Air pollution nanoparticles (airNPs) include combustion-derived nanomaterials, brake wear and tire wear. Nano cerium, for example, is used as a fuel additive in diesel vehicles (inset; representative transmission electron microscopy; bar 50 nm). For more information on airNPs, their abundance and their morphology by transmission electron microscopy, see [17].

A significant increase in the airNP fraction in fuel emissions is an important health issue since animal models identified the efficient translocation of nano-sized manganese oxide particles to the central nervous system (CNS) through the olfactory bulb [20]. The entry route via the olfactory neuroepithelium (i.e., anterograde transport along axons of olfactory sensory neurons across the blood–brain barrier to the CNS) is generally used by certain spherical nanoparticles such as viruses [21–23], explaining both the detection of airNPs in postmortem brains and the associated amyloid neuropathology [17].

Highlighting another important issue, work by Dale et al. demonstrated that the properties of nano ceria differ before and after the combustion process [24]. The consequence of these findings is that comparative nanotoxicological studies of traffic-related airNPs are required that include engineered nanomaterials before and after use—here, collected brake or tire wear and exhaust emissions. Thus, the roadside collection of real-life particle fractions represents an informative tool to investigate the bio-interactions of airNPs. Especially, the comparative analysis of defined pre-use and (inevitably mixed) post-use nanomaterials has the potential to elucidate the mechanistic pathways of potential adverse health effects. Generally, air pollution by traffic is not only composed of airNPs, but represents a plethora of chemicals that is appropriately described as "an exploded pharmacy". These considerations unavoidably augment to a great number of required experiments that take into account (i) the different nanomaterials composing the airNP fraction, (ii) the definition of biophysical NP properties pre- and post-use, (iii) combinations of airNPs with other traffic-related pollutants and (iv) environmental conditions such as urban climate (e.g., temperature).

The current challenge is to develop viable research strategies that allow for comparative low- to high-throughput investigation in order to include a great variety of contributing environmental factors.

1.4. NP-Induced Amyloid Protein Aggregation In Vitro, Cell Culture and the Animal Model Caenorhabditis elegans

In previous work, it was shown that certain nanomaterials, including airNPs, induce amyloid protein aggregation and neurodegeneration using diverse research platforms such as in vitro, cell culture and invertebrate animal models.

In vitro. A seminal paper showed that certain nanomaterials such as copolymer particles, nano cerium, quantum dots and carbon nanotubes enhance the nucleation of protein fibrils of β2-microglobulin in the test tube [25]. These findings corroborated observations that certain surfaces of lipid bilayers, collagen fibers and polysaccharides promote the formation of amyloid fibrils. The concept developed that the unique surface area of nano-sized particles offers a biophysical environment that determines if a nanomaterial catalyzes or inhibits amyloid fibrillation of intrinsically aggregation-prone proteins [26]. The specific interactions between NP surfaces and fibrillation-prone proteins are exploited by engineered nanomaterials such as coated gold NPs that can be used as labels for amyloid fibers in postmortem brains of patients with AD and in other human tissue [27].

Cell culture. The analysis of cultured epithelial cells revealed that nano silica is efficiently taken up into single cells via endocytosis and reaches the cell nucleus within a few seconds [28]. In the cell nucleus, the specific surface area of silica NPs promoted the fibrillation of nuclear proteins to amorphous aggregates that grew over time to amyloid structures [29,30]. Nano-silica-induced intracellular amyloid was located by the amyloid dyes Congo red and ThioflavinT as well as amyloid-specific antibodies [31–33]. Investigation of the proteolytic pathways revealed that the ubiquitin-proteasome system is likewise located in silica-NP-induced amyloid aggregates and possesses proteolytic activity [32].However, this proteolytic activity is not sufficient to dissolve the aggregates. Instead, the nano-silica-induced amyloid seems to be irreversibly insoluble already. Notably, nuclear aggregates generated by silica NPs showed a similar protein composition and analogous biochemical properties to respective pathological amyloid aggregates in neurodegenerative diseases such as AD, PD and Huntington's disease (HD) [29,32].

To summarize, nano silica that likewise constitute airNPs (i.e., brake and tire wear) represent a considerable environmental hazard because in cultured epithelial cells and neurons they induce aberrant protein fibrillation in the cell nucleus that constitutes a pathology resembling the one seen in neurodegenerative aggregation diseases and ataxias [29,32,34]. However, nano-silica-induced amyloid protein fibrillation is not confined to the cell nucleus. The formation of cytoplasmic inclusions, including the aggregation of β-synuclein interacting protein, was observed in neural cell culture and primary cortical or dopaminergic neurons exposed to silica-coated magnetic nanoparticles [35]. The paper highlights the enhanced vulnerability of neurons to the adverse effects of nano silica due to higher levels of reactive oxygen species (ROS), lower proteasomal activity and decreased cell viability. Thus, a unique vulnerability of neurons to nano silica may result from less-efficient detoxification pathways [36,37].

C. elegans. To learn more about the bio-interactions of silica NPs with the nervous system in an organism, further analyses were carried out in the invertebrate animal model *Caenorhabditis elegans*. The nematode roundworm *C. elegans* has a short lifespan of 2–3 weeks and is optimally suited to interrogate NP bio-interactions during a chronic, lowest observed adverse effect level (LOAEL) exposure scenario [38]. Approximately 20,000 genes encode for the nematode's proteins, and the majority (60–80%) of human genes, including disease genes, have a counterpart/homolog in the worm [39,40]. The etiology of neurodegenerative diseases has been extensively investigated using *C. elegans* as a model organism. Consistently, *C. elegans* is used as a tool for the screening of neuroprotective compounds, some of which are running in third phase clinical trials [41].

It was shown that silica NPs enter *C. elegans* effectively via epithelial cells of the reproductive system and the gut [42]. Corroborating the previous results from cultured epithelial and neural cells, the observation of single intestinal cells revealed that silica NPs reach the cell nucleus and induce amyloid in the nucleolus. Concerning the absorption of nutrients, nano silica interferes with the uptake of di- and tripeptides from the intestinal lumen and inhibits their downstream hydrolysis to amino acids [43]. The entire peptide metabolism is disturbed, which results in dwarfism and premature aging of young worms.

In single neurons, protein aggregation, neurodegeneration and defective serotonergic as well as dopaminergic neurosignaling has been observed (Figure 2; [44,45]). This is consistent with the idea that silica NPs interfere with key processes of neuronal function, ranging from nerve impulse transduction to neurotransmitter synthesis and mitochondrial energy production. The neurotoxic endpoints then induce neuromuscular defects such as reduced locomotion and paralysis (Figure 2; [46]).

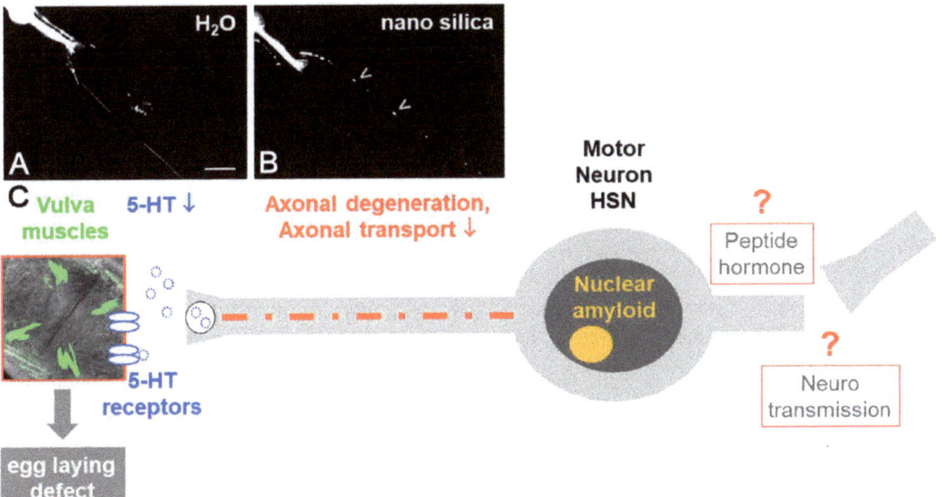

Figure 2. The neurotoxicity of nano silica in the neural system of the nematode *Caenorhabditis elegans*. Representative fluorescence micrographs of a single hermaphrodite-specific neuron (HSN): DsRed reporter worms of the neurotransmitter serotonin (5-hydroxytryptamine, 5-HT) were mock-treated with distilled water (**A**) or exposed to nano silica (**B**). Nano silica induced aggregation of DsRed-5-HT in the axon of the HSN neuron ((**B**), arrowheads). (**C**) Schematic of nano silica effects in *C. elegans*. Silica nanoparticles induce nuclear amyloid in single HSN neurons and protein aggregation on HSN axons. Axonal transport is disturbed and serotonergic neurotransmission at the synapse reduced. This in turn leads to defective vulva muscles and egg-laying defects [44]. Bar, 20 μm.

However, silica NPs not only promote widespread protein aggregation and amyloid formation in wild type (N2) worms. The molecular mechanism of facilitated protein fibrillation by the specific surface area of nano silica was likewise corroborated in *C. elegans* disease models of AD, PD and HD ([42,43] Piechulek and von Mikecz, unpublished observation). In reporter worms of AD, PD and HD, hallmark proteins of the respective neurodegenerative aggregation diseases such as amyloid-β protein, tau protein, α-synuclein and poly-glutamine (polyQ) form aberrant amyloid fibrils in response to exposure to silica NPs.

1.5. Lifetime Exposure of Adult C. elegans to Nanomaterials

Lifespan-resolved nanotoxicology in adult hermaphrodite *C. elegans* involves cultivation in 96-well microtiter plates, where each well represents a specific microenvironment. In the presence of NPs, characteristic aging stigmata were analyzed in differently aged adult nematodes. These included (i) decrease in the rate of locomotion, swimming and pharyngeal pumping, (ii) disorganization of organ morphology (i.e., pharynx, intestine, body wall muscles), (iii) impaired protein homeostasis and increased amyloid formation or (iv) increased neurodegeneration [42,44,47]. Due to the repeated observation that certain nanoparticles reduce the health span of the worms, the concept of an aging dose (AD_{50}) was introduced, which allows for the identification of toxicants that accelerate aging processes in adult *C. elegans*. The AD_{50} enables the detection of nano-sized particles that turn a young worm into an old worm [45]. While the AD_{50} has been established in *C. elegans*, an organism with a short life span, it may also be useful in long-lived individuals. Research across species, including humans, is needed to better understand the role of certain nanomaterials in aging and age-related diseases such as the neurodegenerative diseases AD and PD.

The establishment of lifetime nanotoxicology by the cultivation of adult *C. elegans* in 96-well microtiter plates represents a promising assay that allows for comparative investigations of complex airNP fractions. It enables the interrogation of urgent questions including the biophysical NP properties pre- and post-use, combinations of airNPs with other traffic-related pollutants and additional contributing environmental conditions such as urban climate (e.g., temperature). The latter can be achieved by growing the worms in liquid cultures between 15 and 25 °C.

In order to investigate how silica NPs impact molecular pathways that connect amyloid formation in single cells, global protein aggregation and neuromuscular defects of aging *C. elegans*, mass-spectrometry-based proteomics were used. It was shown that exposure of adult *C. elegans* to nano silica induced the segregation of proteins predominately involved in protein homeostasis and mitochondrial function within an SDS-insoluble aggregome network [44]. Consistently, widespread protein aggregation likewise included the axons of serotonergic hermaphrodite-specific neuron (HSN; Figure 2A). In turn, the impaired axonal transport of the neurotransmitter serotonin to the HSN synapse interrupted the neuromuscular circuit of egg laying and promoted the neural defect internal hatch (Figure 2B; [45]). Since protein aggregation in the HSN and internal hatch was rescued by anti-aging compounds that likewise inhibit amyloid formation, it was concluded that silica NPs cause premature aging in *C. elegans* by a neuropathology driven by imbalanced protein homeostasis.

2. Conclusions and Future Perspectives

There is increasing evidence that certain nanomaterials induce amyloid protein fibrillation in the test tube, cultured epithelial cells and primary neurons as well as diverse tissues of the animal model *C. elegans*, including the neural system. With respect to traffic-related airNPs, most studies have shown an amyloid-induction propensity of nano silica from brake and tire wear or nano ceria, which is used among other things as a diesel fuel additive.

These experimental results coincide with (i) the detection of airNPs in postmortem brains of animals and humans with chronic, long-term exposure to air pollution in metropolitan areas [6], (ii) the detection of airNPs and AD pathology in brains of laboratory mice exposed to traffic-related nanomaterials [13] and (iii) epidemiological data that identifies the proximity of a residence near busy roads as a risk factor for the development of the neurodegenerative disease AD [8].

Urbanization represents a global trend, and it is predicted that in 2050, 68% of the world's population will live in major cities (United Nations, Department of Economic and Social Affairs, https://www.un.org/development/desa/en/news/population/2018-revision-of-world-urbanization-prospects.html, accessed 20 May 2020). To protect the health of these citizens, it seems more than reasonable to investigate potential adverse health effects of air-pollution-related nanomaterials.

How can we reconcile the data from in vitro, cell culture and animal models (i.e., the wet laboratory) with the epidemiological findings? Evidently, innovative and viable research strategies are required. We suggest identifying the molecular pathways of amyloid formation and neurodegeneration by traffic-related nanomaterials in the invertebrate animal model *C. elegans* and verifying the key components of these pathways with whole-genome analyses in humans. The option for medium-throughput, proteomics and lifespan-resolved investigation in *C. elegans* enables the comparative characterization of diverse nanomaterials, their biophysical properties pre- and post-use, diverse environmental conditions and the identification of vulnerable age groups. Key genes or pathways may be directly compared between the *C. elegans* and human genomes of individuals with long-term, chronic exposure to airNPs due to the remarkable homology between the human and *C. elegans* gene repertoires [38–40].

The objective is to distinguish amyloid-promoting nano-sized particles in traffic-related emissions from inactive components and thereby define neurosafe nanomaterials. An obvious question is if the nanomaterials are intrinsically neurotoxic or obtain their potential to induce amyloid formation post-use (i.e., after a high-temperature combustion process). This has consequences for potential mitigation strategies, which may include (i) the replacement of unsafe nanomaterials in tires, brakes or fuel, (ii) the amendment of combustion processes and/or (iii) ultimately, the replacement of fossil fuel combustion in motor traffic. However, rapid mitigation strategies may likewise address the identification of vulnerable target groups for adverse airNP effects in the urban population and the provision of healthier living environments by, for example, the implementation of more green infrastructure in urban planning [48].

Exposure to airNPs could also predispose exposed populations to contracting viral infections and to contracting COVID-19-associated immunopathologies. The severe acute respiratory syndrome coronavirus 2 (SARS-CoV-2) induces neurological complications, and the possible long-term impact for neurological and especially neurodegenerative diseases can only be anticipated [49]. In a worst-case scenario, the common olfactory route of SARS-CoV-2 and airNPs may exacerbate the adverse health effects on the central nervous system. Here, respective investigations are much needed.

Another important issue of future research is the premature onset of neurodegeneration by airNPs. As prenatal air pollution (i.e., diesel exhaust particles) was shown to increase anxiety and impaired cognition in male offspring of mice, the question arises if neurodegenerative defects are imprinted in early development and manifest only later in life [50]. The discussion of if and how aging processes of the neural system originate in prenatal development and the role of air pollutants has just begun.

Author Contributions: Writing—Original Draft Preparation, A.v.M., T.S.; Writing—Review and Editing, A.v.M., T.S.; Visualization, A.v.M.; Funding Acquisition, A.v.M. All authors have read and agreed to the published version of the manuscript.

Funding: This work was funded by the Deutsche Forschungsgemeinschaft, Grant MI 486/10-1.

Acknowledgments: We thank the German Research Foundation DFG (Grant MI 486/10-1) for financial support, Katrin Buder and Peter Hemmerich (FLI-Leibniz Institute for Age Research, Jena) for transmission electron microscopy of CeO_2 nanoparticles and Andrea Scharf (Washington University, St. Louis, USA) for fluorescence micrographs of single *C. elegans* neurons.

Conflicts of Interest: The authors declare no conflict of interest.

References

1. Thal, D.R.; Grinberg, L.T.; Attems, J. Vascular dementia: Different forms of vessel disorders contribute to the development of dementia in the elderly brain. *J. Exp. Gerontol.* **2012**, *47*, 816–824. [CrossRef] [PubMed]
2. Knowles, T.P.; Vendruscolo, M.; Dobson, C.M. The amyloid state and its association with protein misfolding diseases. *Nat. Rev. Mol. Cell Biol.* **2014**, *15*, 384–396. [CrossRef] [PubMed]
3. Dobson, C.M.; Knowles, T.P.J.; Vendruscolo, M. The Amyloid Phenomenon and Its Significance in Biology and Medicine. *Cold Spring Harb. Perspect. Biol.* **2020**, *12*, a033878. [CrossRef] [PubMed]
4. Ball, N.; Teo, W.P.; Chandra, S.; Chapman, J. Parkinson's Disease and the Environment. *Front. Neurol.* **2019**, *10*, 218. [CrossRef] [PubMed]
5. Lelieveld, J.; Evans, J.S.; Fnais, M.; Giannadaki, D.; Pozzer, A. The contribution of outdoor air pollution sources to premature mortality on a global scale. *Nature* **2015**, *525*, 367–371. [CrossRef] [PubMed]
6. Calderón-Garcidueñas, L.; Reynoso-Robles, R.; González-Maciel, A. Combustion and friction-derived nanoparticles and industrial-sourced nanoparticles: The culprit of Alzheimer and Parkinson's diseases. *Environ. Res.* **2019**, *176*, 108574. [CrossRef]
7. Paul, K.C.; Haan, M.; Mayeda, E.R.; Ritz, B.R. Ambient Air Pollution, Noise, and Late-Life Cognitive Decline and Dementia Risk. *Annu Rev. Public Health* **2019**, *40*, 203–220. [CrossRef]
8. Chen, H.; Kwong, J.C.; Copes, R.; Tu, K.; Villeneuve, P.J.; van Donkelaar, A.; Hystad, P.; Martin, R.V.; Murray, B.J.; Jessiman, B.; et al. Living near major roads and the incidence of dementia, Parkinson's disease, and multiple sclerosis: A population-based cohort study. *Lancet* **2017**, *389*, 718–726. [CrossRef]
9. Jung, C.R.; Lin, Y.T.; Hwang, B.F. Ozone, particulate matter, and newly diagnosed Alzheimer's disease: A population-based cohort study in Taiwan. *J. Alzheimers Dis.* **2015**, *44*, 573–584. [CrossRef]
10. Ranft, U.; Schikowski, T.; Sugiri, D.; Krutmann, J.; Krämer, U. Long-term exposure to traffic-related particulate matter impairs cognitive function in the elderly. *Environ. Res.* **2009**, *109*, 1004–1011. [CrossRef]
11. Levesque, S.; Surace, M.J.; McDonald, J.; Block, M.L. Air pollution and the brain: Subchronic diesel exhaust exposure causes neuroinflammation and elevates early markers of neurodegenerative disease. *J. Neuroinflamm.* **2011**, *8*, 105. [CrossRef] [PubMed]
12. Calderón-Garcidueñas, L.; Mora-Tiscareno, A.; Styner, M.; Gomez-Garza, G.; Zhu, H.; Torres-Jardon, R.; Carlos, E.; Solorio-Lopez, E.; Medina-Cortina, H.; Kavanaugh, M.; et al. White matter hyperintensities, systemic inflammation, brain growth, and cognitive functions in children exposed to air pollution. *J. Alzheimers Dis.* **2012**, *31*, 183–191. [CrossRef] [PubMed]
13. Calderón-Garcidueñas, L.; Herrera-Soto, A.; Jury, N.; Maher, B.A.; González-Maciel, A.; Reynoso-Robles, R.; Ruiz-Rudolph, P.; van Zundert, B.; Varela-Nallar, L. Reduced repressive epigenetic marks, increased DNA damage and Alzheimer's disease hallmarks in the brain of humans and mice exposed to particulate urban air pollution. *Environ. Res.* **2020**, *183*, 109226. [CrossRef] [PubMed]
14. Schikowski, T.; Altug, H. The role of air pollution in cognitive impairment and decline. *Neurochem. Int.* **2020**, *136*, 104708. [CrossRef] [PubMed]
15. Hoag, H. How cities can beat the heat. *Nature* **2015**, *524*, 402–404. [CrossRef]
16. Mann, M.E.; Rahmstorf, S.; Kornhuber, K.; Steinman, B.A.; Miller, S.K.; Petri, S.; Coumou, D. Projected changes in persistent extreme summer weather events: The role of quasi-resonant amplification. *Sci. Adv.* **2018**, *4*, eaat3272. [CrossRef]
17. Gonet, T.; Maher, B.A. Airborne, Vehicle-Derived Fe-Bearing Nanoparticles in the Urban Environment: A Review. *Environ. Sci. Technol.* **2019**, *53*, 9970–9991. [CrossRef]
18. Pant, P.; Harrison, R.M. Estimation of the contribution of road traffic emissions to particulate matter concentrations from field measurements. *Rev. Atmos. Environ.* **2013**, *77*, 78–97. [CrossRef]
19. Zhang, J.; Nazarenko, Y.; Zhang, L.; Calderon, L.; Lee, K.B.; Garfunkel, E.; Schwander, S.; Tetley, T.D.; Chung, K.F.; Porter, A.E.; et al. Impacts of a nanosized ceria additive on diesel engine emissions of particulate and gaseous pollutants. *Environ. Sci. Technol.* **2013**, *47*, 13077–13085. [CrossRef]
20. Elder, A.; Gelein, R.; Silva, V.; Feikert, T.; Opanashuk, L.; Carter, J.; Potter, R.; Maynard, A.; Ito, Y.; Finkelstein, J.; et al. Translocation of inhaled ultrafine manganese oxide particles to the central nervous system. *Environ. Health Perspect.* **2006**, *114*, 1172–1178. [CrossRef]

21. Netland, J.; Meyerholz, D.K.; Moore, S.; Cassell, M.; Perlman, S. Severe acute respiratory syndrome coronavirus infection causes neuronal death in the absence of encephalitis in mice transgenic for human ACE2. *J. Virol.* **2008**, *82*, 7264–7275. [CrossRef] [PubMed]
22. Durrant, D.M.; Ghosh, S.; Klein, R.S. The Olfactory Bulb: An Immunosensory Effector Organ during Neurotropic Viral Infections. *ACS Chem. Neurosci.* **2016**, *7*, 464–469. [CrossRef] [PubMed]
23. Dubé, M.; Le Coupanec, A.; Wong, A.H.M.; Rini, J.M.; Desforges, M.; Talbot, P.J. Axonal Transport Enables Neuron-to-Neuron Propagation of Human Coronavirus OC43. *J. Virol.* **2018**, *92*, e00404–e00418. [CrossRef]
24. Dale, J.G.; Cox, S.S.; Vance, M.E.; Marr, L.C.; Hochella, M.F., Jr. Transformation of Cerium Oxide Nanoparticles from a Diesel Fuel Additive during Combustion in a Diesel Engine. *Environ. Sci. Technol.* **2017**, *51*, 1973–1980. [CrossRef]
25. Linse, S.; Cabaleiro-Lago, C.; Xue, W.F.; Lynch, I.; Lindman, S.; Thulin, E.; Radford, S.E.; Dawson, K.A. Nucleation of protein fibrillation by nanoparticles. *Proc. Natl. Acad. Sci. USA* **2007**, *104*, 8691–8696. [CrossRef]
26. John, T.; Gladytz, A.; Kubeil, C.; Martin, L.L.; Risselada, H.J.; Abel, B. Impact of nanoparticles on amyloid peptide and protein aggregation: A review with a focus on gold nanoparticles. *Nanoscale* **2018**, *10*, 20894–20913. [CrossRef]
27. Cendrowska, U.; Silva, P.J.; Ait-Bouziad, N.; Müller, M.; Guven, Z.P.; Vieweg, S.; Chiki, A.; Radamaker, L.; Kumar, S.T.; Fändrich, M.; et al. Unraveling the complexity of amyloid polymorphism using gold nanoparticles and cryo-EM. *Proc. Natl. Acad. Sci. USA* **2020**, *117*, 6866–6874. [CrossRef]
28. Hemmerich, P.H.; von Mikecz, A. Defining the subcellular interface of nanoparticles by live-cell imaging. *PLoS ONE* **2013**, *8*, e62018. [CrossRef]
29. Chen, M.; von Mikecz, A. Formation of nucleoplasmic protein aggregates impairs nuclear function in response to SiO_2 nanoparticles. *Exp. Cell Res.* **2005**, *305*, 51–62. [CrossRef]
30. Chen, M.; von Mikecz, A. Nanoparticle-induced cell culture models for degenerative protein aggregation diseases. *Inhal. Toxicol.* **2009**, *21*, 110–114.
31. LeVine, H. Thiofla0vine T interaction with synthetic Alzheimer's disease beta-amyloid peptides: Detection of amyloid aggregation in solution. *Protein Sci.* **1993**, *2*, 404–410. [CrossRef] [PubMed]
32. Chen, M.; Singer, L.; Scharf, A.; von Mikecz, A. Nuclear polyglutamine-containing protein aggregates as active proteolytic centers. *J. Cell Biol.* **2008**, *180*, 697–704. [CrossRef] [PubMed]
33. Arnhold, F.; Gührs, K.H.; von Mikecz, A. Amyloid domains in the cell nucleus controlled by nucleoskeletal protein lamin B1 reveal a new pathway of mercury neurotoxicity. *PeerJ* **2015**, *3*, e754. [CrossRef] [PubMed]
34. Ross, C.A.; Poirier, M.A. Protein aggregation and neurodegenerative disease. *Nat. Med.* **2004**, *10*, S10–S17. [CrossRef]
35. Phukan, G.; Shin, T.H.; Shim, J.S.; Paik, M.J.; Lee, J.K.; Choi, S.; Kim, Y.M.; Kang, S.H.; Kim, H.S.; Kang, Y.; et al. Silica-coated magnetic nanoparticles impair proteasome activity and increase the formation of cytoplasmic inclusion bodies in vitro. *Sci. Rep.* **2016**, *6*, 29095. [CrossRef]
36. Wang, X.; Michaelis, E.K. Selective neuronal vulnerability to oxidative stress in the brain. *Front. Aging Neurosci.* **2010**, *2*, 12. [CrossRef]
37. Reckziegel, P.; Chen, P.; Caito, S.; Gubert, P.; Soares, F.A.; Fachinetto, R.; Aschner, M. Extracellular dopamine and alterations on dopamine transporter are related to reserpine toxicity in Caenorhabditis elegans. *Arch. Toxicol.* **2016**, *90*, 633–645. [CrossRef]
38. von Mikecz, A. Lifetime eco-nanotoxicology in an adult organism: Where and when is the invertebrate *C. elegans* vulnerable? *Environ. Sci. Nano* **2018**, *5*, 616–622. [CrossRef]
39. *C. elegans* Sequencing Consortium. Genome sequence of the nematode *C. elegans*: A platform for investigating biology. *Science* **1998**, *282*, 2012–2018. [CrossRef]
40. Kaletta, T.; Hengartner, M.O. Finding function in novel targets: *C. elegans* as a model organism. *Nat. Rev. Drug Discov.* **2006**, *5*, 387–398. [CrossRef]
41. Wischik, C.M.; Staff, R.T.; Wischik, D.J.; Bentham, P.; Murray, A.D.; Storey, J.M.; Kook, K.A.; Harrington, C.R. Tau aggregation inhibitor therapy: An exploratory phase 2 study in mild or moderate Alzheimer's disease. *J. Alzheimers Dis.* **2015**, *44*, 705–720. [CrossRef] [PubMed]
42. Scharf, A.; Piechulek, A.; von Mikecz, A. Effect of nanoparticles on the biochemical and behavioral aging phenotype of the nematode Caenorhabditis elegans. *ACS Nano* **2013**, *7*, 10695–10703. [CrossRef] [PubMed]

43. Piechulek, A.; Berwanger, L.C.; von Mikecz, A. Silica nanoparticles disrupt OPT-2/PEP-2-dependent trafficking of nutrient peptides in the intestinal epithelium. *Nanotoxicology* **2019**, *13*, 1133–1148. [CrossRef]
44. Scharf, A.; Gührs, K.H.; von Mikecz, A. Anti-amyloid compounds protect from silica nanoparticle-induced neurotoxicity in the nematode C. elegans. *Nanotoxicology* **2016**, *10*, 426–435. [CrossRef] [PubMed]
45. Piechulek, A.; von Mikecz, A. Aging by pollutants: Introducing the aging dose (AD) 50. *Environ. Sci. Eur.* **2019**, *31*, 23. [CrossRef]
46. Piechulek, A.; von Mikecz, A. Life span-resolved nanotoxicology enables identification of age-associated neuromuscular vulnerabilities in the nematode Caenorhabditis elegans. *Environ. Pollut.* **2018**, *233*, 1095–1103. [CrossRef]
47. Walther, D.M.; Kasturi, P.; Zheng, M.; Pinkert, S.; Vecchi, G.; Ciryam, P.; Morimoto, R.I.; Dobson, C.M.; Vendruscolo, M.; Mann, M.; et al. Widespread Proteome Remodeling and Aggregation in Aging, *C. elegans*. *Cell* **2015**, *161*, 919–932. [CrossRef]
48. Wang, H.; Maher, B.A.; Ahmed, I.A.; Davison, B. Efficient Removal of Ultrafine Particles from Diesel Exhaust by Selected Tree Species: Implications for Roadside Planting for Improving the Quality of Urban Air. *Environ. Sci. Technol.* **2019**, *53*, 6906–6916. [CrossRef]
49. DeFelice, F.G.; Tovar-Moll, F.; Moll, J.; Munoz, D.P.; Ferreira, S.T. Severe acute respiratory syndrome coronavirus 2 (SARS-CoV-2) and the central nervous system. *Trends Neurosci.* **2020**, *43*, 355–357. [CrossRef]
50. Bolton, J.L.; Huff, N.C.; Smith, S.H.; Mason, S.N.; Foster, W.M.; Auten, R.L.; Bilbo, S.D. Maternal stress and effects of prenatal air pollution on offspring mental health outcomes in mice. *Environ. Health Perspect.* **2013**, *121*, 1075–1082. [CrossRef]

Review

Nano Meets Micro-Translational Nanotechnology in Medicine: Nano-Based Applications for Early Tumor Detection and Therapy

Svenja Siemer [1], Désirée Wünsch [1], Aya Khamis [1], Qiang Lu [1], Arnaud Scherberich [2], Miriam Filippi [2], Marie Pierre Krafft [3], Jan Hagemann [1], Carsten Weiss [4], Guo-Bin Ding [5], Roland H. Stauber [5,1,*] and Alena Gribko [1,*]

1 Nanobiomedicine Department, University Medical Center Mainz/ENT, Langenbeckstrasse 1, 55131 Mainz, Germany; svenja.siemer@uni-mainz.de (S.S.); wuensch@uni-mainz.de (D.W.); ayakhamis@uni-mainz.de (A.K.); qianglu@uni-mainz.de (Q.L.); jan.hagemann@unimedizin-mainz.de (J.H.)

2 Laboratory of Tissue Engineering, Universitätspital Basel, Hebelstrasse 20, CH-4031 Basel, Switzerland; arnaud.scherberich@unibas.ch (A.S.); miriam.filippi@usb.ch (M.F.)

3 Institut Charles Sadron (CNRS), University of Strasbourg, 23 rue du Loess, 67034 Strasbourg Cedex, France; marie-pierre.krafft@ics-cnrs.unistra.fr

4 Institute of Biological and Chemical Systems-Biological Information Processing (IBCS-BIP), Postfach 3640, 76021 Karlsruhe, Germany; carsten.weiss@kit.edu

5 Institute for Biotechnology, Shanxi University, No. 92 Wucheng Road, 030006 Taiyuan, China; dinggb2012@sxu.edu.cn

* Correspondence: roland.stauber@unimedizin-mainz.de (R.H.S.); algribko@uni-mainz.de (A.G.); Tel.: +49-6131-176030 (A.G.)

Received: 10 January 2020; Accepted: 15 February 2020; Published: 22 February 2020

Abstract: Nanomaterials have great potential for the prevention and treatment of cancer. Circulating tumor cells (CTCs) are cancer cells of solid tumor origin entering the peripheral blood after detachment from a primary tumor. The occurrence and circulation of CTCs are accepted as a prerequisite for the formation of metastases, which is the major cause of cancer-associated deaths. Due to their clinical significance CTCs are intensively discussed to be used as liquid biopsy for early diagnosis and prognosis of cancer. However, there are substantial challenges for the clinical use of CTCs based on their extreme rarity and heterogeneous biology. Therefore, methods for effective isolation and detection of CTCs are urgently needed. With the rapid development of nanotechnology and its wide applications in the biomedical field, researchers have designed various nano-sized systems with the capability of CTCs detection, isolation, and CTCs-targeted cancer therapy. In the present review, we summarize the underlying mechanisms of CTC-associated tumor metastasis, and give detailed information about the unique properties of CTCs that can be harnessed for their effective analytical detection and enrichment. Furthermore, we want to give an overview of representative nano-systems for CTC isolation, and highlight recent achievements in microfluidics and lab-on-a-chip technologies. We also emphasize the recent advances in nano-based CTCs-targeted cancer therapy. We conclude by critically discussing recent CTC-based nano-systems with high therapeutic and diagnostic potential as well as their biocompatibility as a practical example of applied nanotechnology.

Keywords: biocompatibility; circulating tumor cells; metastasis; microbubbles; nanomedicine; nanotechnology

1. Introduction

The application of engineered nanomaterials (NMs) in technical products is steadily growing in biotechnology and biomedicine [1]. Nanomedicine, i.e., the medical application of nanotechnology, is expected to play a vital role in early tumor detection and cancer treatment. The primary cause of cancer morbidity and mortality is cancer metastasis. It is estimated that about 90% of cancer deaths are caused by metastasis [2–5]. This process is determined as the dissemination of cancer cells from primary tumors to surrounding tissues and to distant organs, which is also known as the invasion-metastasis cascade. One necessary step in distant metastasis is the transport of tumor cells through the blood system, but detailed molecular mechanisms underlying tumor metastasis still remain unclear [6,7]. Detached cancer cells of solid tumor origin from primary tumor which intravasate into the peripheral blood system and circulate in the body are called circulating tumor cells (CTCs). Only a small number of CTCs are able to evade immune attack and extravagate during the circulation at distant capillary beds and seed the growth of a secondary tumor [8]. Consequently, CTCs play an important role as part of a 'liquid biopsy' which can offer important information on prediction of cancer progression and survival after specific treatment without surgery [9,10]. The analysis of CTCs includes characterization, determination and enumeration of CTCs. Comparison studies of enumerating CTCs before and after resection open up the possibility for monitoring therapeutic response. Moreover, the enumeration of CTCs also represents an attractive biomarker for predicting the possibility of tumor recurrence [3]. Furthermore, the number of detected CTCs usually correlates with the progression of cancer disease resulting in further information about tumor burden and recurrence [11–13]. Additionally, cultivation of isolated patients-derived CTCs can be used for drug resistance analyses and also for the development of personalized anti-cancer agents (Figure 1) [11,14].

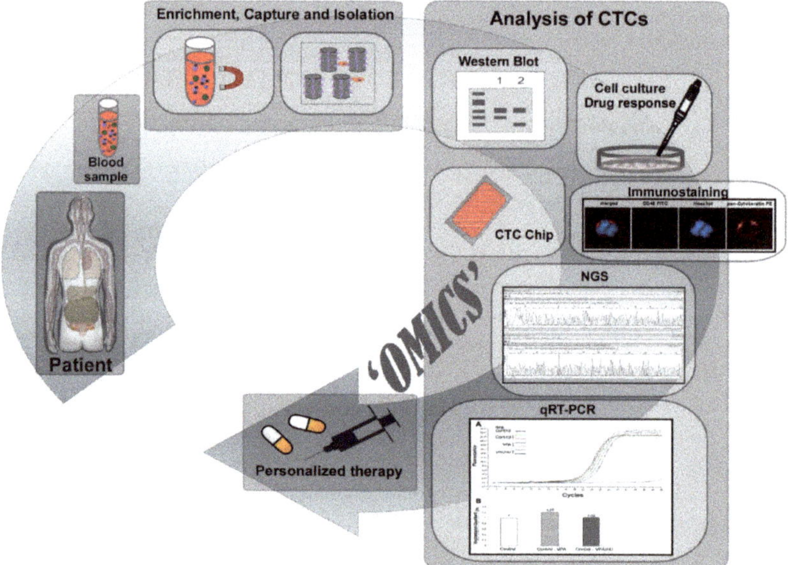

Figure 1. Overview of CTCs analysis of patients-derived CTCs including drug resistance detection and personalized therapy: Patients' blood sample is screened and potential CTCs are captured and isolated by different isolation methods. Potential CTCs can be determined and used for further analysis to develop personalized medicine.

During the early stages of tumorigenesis the determination of the existence of CTCs in blood samples of patients is a significant biomarker for early cancer detection [15]. Moreover, CTCs have been detected in many cancer types including breast [16], colon [17], lung [18], melanoma [2], ovarian [19] and prostate cancers [20]. Nevertheless, because of CTCs rarity and their property to move as individual cells or as multi-cellular clumps, their capture and detection are extremely challenging. For example, in 1 mL blood sample of an early stage cancer patient can be detected approximately five billion red blood cells, ten million white blood cells and as few as one CTC [21,22]. Because of biological and molecular changes of CTCs during the epithelial-to-mesenchymal transition (EMT), the circulating cell population is heterogenic and requires the ability to handle a very small number of cells for efficient isolation methods [23].

Over the last few decades, nanoscale materials have been used in a wide range of areas such as electronics, energy conversion, catalysis, storage and medicine. The variety of these advanced nanoscale materials includes metal, metal oxide, semiconductor, polymeric NMs and microbubbles (MBs) [21,24]. Excellent contributions to clinical medicine were made by NMs since they possess some attractive properties related to their size, shape and surface characteristics [25]. Due to the nanoscale effect, NMs have surpassing structural and functional properties that are typically different from either bulk materials or discrete molecules. Although CTCs were already discovered in 1869 by the Australian researcher Thomas Ashworth, only during the last two decades a large number of important advancements have been made in the field of CTC isolation and detection techniques [26]. In recent years, nanomedicine (i.e., here the use of NMs for CTC detection and isolation) has been playing an increasingly important role in CTC detection and more than 100 companies are providing CTC related products and services [27]. The specificity of CTC recognition could be significantly improved by conjugation of NMs with targeting ligands. It has been shown that functionalized substrates and captured CTCs exhibit improved ligand-antigen binding. Nanostructured substrates demonstrate enhanced local topographic interactions that lead to enhanced cell capture affinity. NMs can also be used as drug delivery systems for CTC-targeted drug delivery and cancer treatment [28]. Moreover, NMs have a large surface-to-volume ratio that endows them with a high cellular binding affinity in the complex blood matrix. Additionally, ligand coating of nanostructures can be prepared with much higher density improving binding affinity in comparison to micro- and macrostructures. A manipulation of NMs gives them the ability of multiplexed detection and targeting, which are crucial to approach the heterogeneous problem of CTCs [21]. Furthermore, the use of microfluidic chips as cost-effective, miniaturized and efficiency improved applications for the enrichment and detection of CTCs obtain better performance with nanostructured substrates. When targeted MBs are used, CTCs can readily be separated by simple flotation or gentle centrifugation, preserving cell viability for culture and also providing theranostic capacities.

In this review, we will provide an overview of current CTC enrichment strategies and clarify the relationship between CTCs and tumor metastasis. We will discuss the interactions of nanoparticles (NPs) and MBs with biomolecules such as proteins in biological media, and what consequences this may have on detection and isolation strategies. Some CTC detection and analysis methods will briefly be discussed as a guide for the development of potential clinical diagnostic platforms. Literature on in vitro NPs-based CTCs enrichment systems which have drawn extensive attention due to their clinical potential will also be elaborated in this review. Besides our focus on the "nano", we will also elucidate the "micro" including complementary microfluidic and lab-on-a-chip technologies for simultaneous CTCs enrichment and analysis. Last but not least, we will summarize the research progress of the development of robust nanosystems and MBs for CTC-targeted cancer therapy.

2. The Metastatic Process

As previously mentioned, metastasis is a multi-step process including the spread of cancer cells from primary tumor to distant organs by intravasation into the circulating system. These cells are often called circulating tumor stem cells due to their stem-like properties [29]. CTCs are involved in the process of EMT during early steps of the metastatic cascade [30]. This process is involved in breaking up the cell-to-cell contact by downregulation of various cell adhesion molecules (for example E-cadherin), or epithelial antigens (like epithelial cell adhesion molecule, EpCAM). Due to specific signaling molecules (*Wnt/β-catenin*, *FGF* or *TGFβ1/BMP*) EMT induces cell migration and development of mesenchymal-like cells [31]. During EMT CTCs detach from primary tumor, loose their epithelial character by downregulation of EpCAM, infiltrate the blood circulation system and migrate into distant site of future metastasis [7,32]. At distant sites, CTCs interact with the local microenvironment which leads to its adaption via developmental and self-renewal signaling pathways, like Hh, Wnt and Notch. These signaling pathways are responsible for the proliferation and finally for the forming of metastases [33]. Subsequently, CTCs have to recover their epithelial characteristic to resettle in the target organ. This process is called mesenchymal-to-epithelial transition (MET) and is a reverse process of EMT [34]. The process of MET at distant sites is not fully understood and it is also not known how many factors are responsible for MET activation. Banyard et al. demonstrated evidence for spontaneous MET process in an in vivo model [35]. This researcher group selected and expanded metastatic cancer cells that survived in the lymph node microenvironment of mice bearing human prostate tumors. The progression of lymphatic cancer cells demonstrated the existence of epithelial like cells as a result of MET. There are also evidences for MET activation after switching off EMT transcription factors, such as Twist 1, and silencing the EMT inducer Prrx1 to prevent further EMT and allow CTCs to migrate into distant sites of future metastasis [35–38].

The process of EMT is important for the initiation of a stem cell phenotype which displays some characteristics. During the metastasizing process developed characteristics such as high invasiveness, self-renewal ability and resistance to apoptosis and therapy, are used as biomarkers for detection and isolation of CTCs [39,40]. Specific biomarkers are essential for most biological detection methods. Cancer biomarkers are the measurable molecular changes between normal and cancerous tissues of patients. Each cancer type has specific pathological evolution and molecular characteristics. Consequently, for further applications in CTC capture and isolation the identification of these biomarkers is crucial [41,42]. For example, CTCs are commonly described to express epithelial markers like EpCAM and cytokeratins (CKs), and to be nucleated (identified by staining with a nuclear dye such as DAPI, 4′, 6-diamidino-2-phenylindole). Moreover, CTCs do not express cell surface marker CD45 which is specific for white blood cells [43–46]. In summary, it can be declared that positive results in CTC specific detection can be obtained by using a variety of epithelial-, mesenchymal-, and stem cell markers. Additionally, in order to determine and eliminate debatable cells 'negative markers' could be used for CTC detection. These markers include for example platelet marker CD61, CD45 and apoptosis marker M30 [43–46]. Some of these key biomarkers are illustrated in Figure 2.

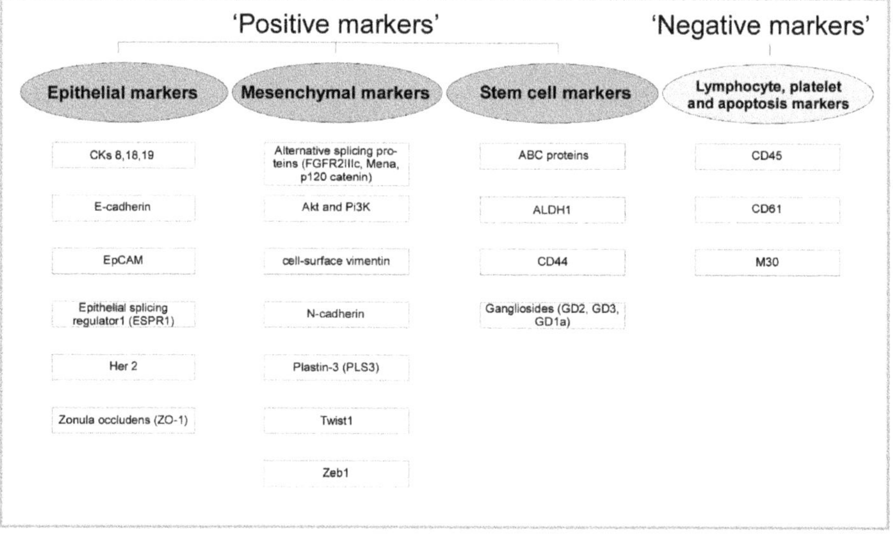

Figure 2. Illustration of CTCs detection specific key biomarkers.

3. Use of Nanomaterials in Tumor Detection and Isolation

A high level of sensitivity as well as specificity due to the extreme rarity and heterogeneous phenotype of CTCs in the blood system are necessary for an effective capture and accurate identification of CTCs [28]. Current CTC enrichment methods are based on either biological features (cell surface protein expression, invasive capacity, and viability) or physical properties (size, density, deformability and electrical charge) [47]. Commonly used isolation methods based on physical properties include membrane filtration, flotation, density gradient centrifugation and microchip-based capture platforms. These methods are fast, simple and label-free, but unfortunately less specific. Accordingly, 'physical-methods' are usually combined with the antibody-labeled biological method. Examples for biological characteristics-dependent isolation methods include immunomagnetic separation, buoyancy-based separation, substrate- and microchip-based capture platforms [21].

As mentioned, a high level of specificity and sensitivity is important for capture and identification of CTCs. NMs could largely improve the sensitivity and efficiency of CTCs enrichment, isolation and detection. Correspondingly, the unique properties of NMs can be used to accelerate detection and overcome some limitations in CTC detection [28]. It is necessary to understand the interaction of NPs with cells, tissues and organisms in detail to reach safe application of NMs as diagnostic devices in cancer therapy [48].

To understand changes of NMs in complex physiological or natural environments an extensive understanding about the behavioral and physico-chemical properties of NMs is urgently needed [1]. Due to a high surface-to-volume ratio, NPs interact with (bio)-molecules upon contact with biological and abiotic environment and form the so-called (bio)-molecular corona. In complex biological environments including simple and higher organisms this (bio)-molecular corona is formed spontaneously like an adsorption layer on the NP. Additionally, this adsorption layer plays an important role in the interaction of NPs with organisms and control of their physiological responses. The biomolecular adsorption is mediated by different properties of NPs, including composition, shape, size, surface charge and surface functionalization [49–52]. The protein absorption to NP surfaces is known as the 'Vroman effect' and thus was postulated in the pioneering work by Vroman [53]. This effect is described as dynamic change of protein corona composition by adsorption and desorption. This means that in a blood sample containing thousands of different proteins, abundant proteins will be desorbed from NP surface and replaced by rare ones with a higher affinity which leads to a constant level of adsorbed proteins [52,54,55].

In 2007, for the first time the term 'protein corona' was introduced to the NP community by Cedervall [54]. The term 'hard protein corona' was described as a strong bound layer of biomolecules, representing an analytically approachable protein/biomolecule signature of NP in a determined environment [1,49,55]. Some models additionally describe a 'soft protein corona' around the 'hard protein corona' which is described as a rapidly exchanging and highly complex biomolecule layer without direct contact to the NPs [1,49,54,56,57]. However, the presence of this 'soft protein corona' (also called 'soft corona cloud') and its importance at the nano-bio interface are not yet fully affirmed. Moreover, in the context of biology and medicine unspecific ligand-receptor interactions have been discussed and no differences between 'soft' and 'hard' ligand-receptor interactions were made. Therefore, it is recommended to term the analytically approachable NP-protein complex as 'protein corona', because the terming 'soft' versus 'hard' corona does not take into account all types of coronas and does not assist in resolving pressing scientific questions [1].

3.1. Magnetic Nanoparticle-Based System

Magnetic separation using magnetic NPs (MNPs) is principally used for the isolation of CTCs. Assembled in an organic or inorganic matrix, dispersed antibody-labelled MNPs or MNPs clusters are bound to CTCs. Due to this composition, cells can be separated via an external magnetic field [21,22]. During the presence of an external magnetic field a magnetic moment is exhibited by the most commonly used MNPs such as cobalt, chromium, iron and also their oxides [58]. The magnetic response of iron oxide MNPs can be ferromagnetic or superparamagnetic depending on the particle shape and size. Moreover, this type MNPs presents chemically stable and biocompatible features [59]. In comparison to iron oxide MNPs, ferromagnetic NPs have a remnant magnetization after removal of the external magnetic field. Moreover, ferromagnetic NPs demonstrate poor stability leading to aggregation in aqueous media so that these particles are not used for cell isolation. Consequently, superparamagnetic NPs (SMNPs) or clusters composed of SMNPs are suitable for cell isolation because of thermal fluctuations [21,22,60]. Additionally, the surface of SMNPs is often modified by coating or grafting with surfactants, polymers, (polyethylene glycol-PEG), polypeptides or hydrophilic inorganic materials (silica and gold) [22].

The most commonly used magnetic system for CTCs isolation is the 'Food and Drug Administration' (FDA)-approved Cell Search system (Menarini Silicon Biosystems Inc, Huntington Valley, PA, USA) which is considered to be the gold standard. This system enriches CTCs using iron NPs (ferrofluid particles) linked with anti-EpCAM antibodies [61] (Figure 3A). The CellSearch system is primarily designed for the enumeration of CTCs with an epithelial origin expressing EpCAM and keratin. Due to the proportional correlation of the magnetic force and the number of bound NPs [62], cells can be selectively enriched by making use of the fact that NPs bound cells are isolated faster than free NPs in a solution under the same external permanent magnet field. This process is separated into two steps and also two different instruments: Autoprep is responsible for CTC capture and immunostaining, and CellTracks Analyzer evaluates the immunofluorescent-stained cells by a semi-automated fluorescence microscope. Further immunofluorescence staining with anti-keratin and anti-CD45 can increase the specificity of selected cells [58,63]. Although CellSearch represents a clinically validated method for CTC isolation, this system has to overcome large limitations including the dependence on cells expressing EpCAM and the fact that only a very small proportion of CTCs in the blood sample of a patient can be detected in a limited interval of time. The process of EMT and the accompanied downregulation of epithelial markers like EpCAM have already been discussed above [58].

Schüling et al. demonstrated aptamers as a suitable alternative to antibodies for whole cell detection with many advantages. High binding specificity is one of the key advantages of aptamer used applications. Despite comparable affinities to antibodies, aptamers present a limited affinity to negatively charged targets. Unfortunately, developed aptamer-based lateral flow assays are not commercially available at the moment because of missing integration in new nano-sized technologies [64].

The magnetic activated cell sorting (MACS, Miltenyi Biotec, Bergisch-Gladbach, Germany) represents a variation of the magnetic isolation method. MACS uses superparamagnetic Fe NPs combined with a magnetized steel wool column as a special feature in comparison to another magnetic-based isolation system. Cells can be eluted from the column by removing the column from the external magnetic field (Figure 3B). By using a combination of magnetic beads coupled with various antibodies and also the possibility of labeling cells with fluorescent antibodies, this technique describes a large advantage due to a direct enrichment and evaluation of captured cells without further detaching or staining procedures [65].

Another method using more than one antibody for the magnetic enrichment of CTCs is the AdnaTest (AdnaGen AG, Langenhagen, Germany). AdnaTest allows the immunomagnetic enrichment of CTCs via epithelial and tumor-specific antigens (Figure 3C) by making use of different magnetic microbeads, such as the superparamagnetic DynaBeads. This mixture of magnetic beads is simultaneously conjugated to antibodies against EpCAM and tumor-associated antigens for labeling of CTCs in peripheral blood. Next, labeled cells are lysed, mRNA is extracted from captured cells and transcribed into cDNA. The analysis of the CTC gene expression can be made by a multiplex polymerase chain reaction (PCR) [66,67]. In comparison to CellSearch, AdnaTest exhibits improved enrichment efficiency due to the usage of two antibodies and the size of magnetic particles.

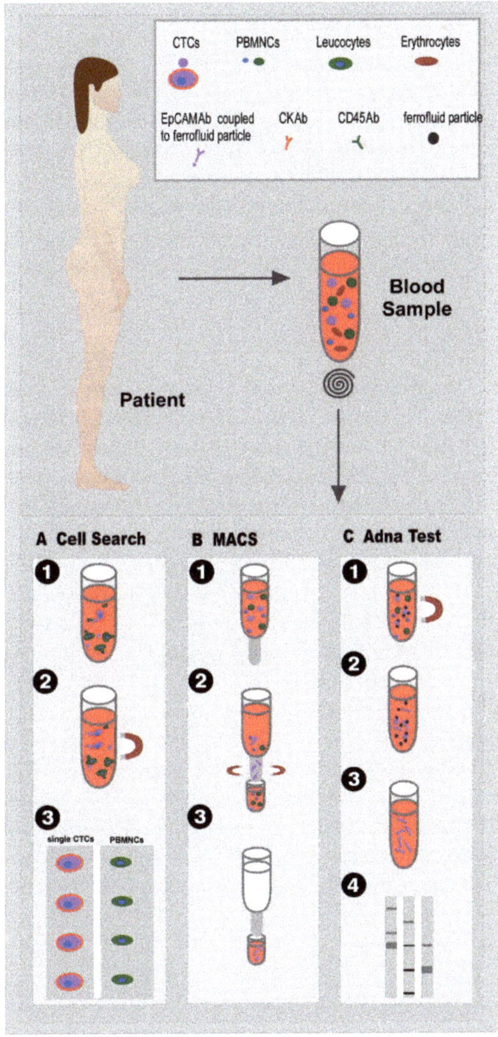

Figure 3. Illustration of CTC isolation methods by magnetic separation using MNPs: (**A**) The *CellSearch* system includes the enrichment of CTCs with ferrofluid particles linked with anti-EpCAM antibodies, magnetic separation of labeled cells and evaluation by immunofluorescent staining. (**B**) The principle of magnetic activated cell sorting (MACS) by using superparamagnetic Fe NPs within a magnetized steel wool column. (**C**) The process of AdnaTest describes the immunomagnetic enrichment of CTCs via epithelial and tumor-specific antigens. Potential CTCs are separated from peripheral blood mononuclear cells (PBMCs) and lysed in order to analyze the CTC gene expression via multiplex PCR.

These three methods represent positive selection strategies for the specific isolation of CTCs out of a bulk of other cells. One large limitation of positive CTC selection is the described necessity of the expression of targeted markers on the surface of cells. A possible solution to overcome this hurdle is the use of negative depletion strategies with magnetic beads. For negative depletion a two-step procedure was suggested including lysis of red blood cells and removing white blood cells by labeling with CD45-specific MNPs. In summary, it remains a great challenge to efficiently capture CTCs, reduce the great number of normal blood cells in a sample, and protect rare CTCs from damage during lysis and different washing steps [22].

For implementation of standardized CTC detection methods in daily clinical routine it is indispensable to compare different methods and determine the efficiency of the technology. Andreopoulou et al. compared CellSearch system and AdnaTest to evaluate CTC detection in peripheral blood samples of 55 metastatic breast cancer patients (2012). In this study the *CellSearch* system demonstrated 26 of 55 patients as CTC positive in comparison to 29 of 55 patients detected as CTC positive by using AdnaTest. Consequently, the detection efficiency of CTCs in metastatic breast cancer patients of both compared techniques is comparable. However, more studies are urgently needed to compare the CTC detection efficiency of described positive selection methods by using the same biological samples.

3.2. Fluorescence-Based Detection by Using Quantum Dots

Fluorescence detection methods take also an important part in leading techniques for CTC detection. For this reason, the use of organic dyes as imaging agents belongs to the standard, although their use is limited by low signal intensity, spectral overlapping, the necessity of multiple light sources to excite different fluorophores in mixed detection and photobleaching [21,68]. Examples for cytometric methods are immunohistochemical staining, flow cytometry and spectroscopic detection. The advantage of cytometric methods is the possibility to further analyze detected cells, if cell lysis is not necessary for former procedures, and to examine the cell morphology, if cells are reported microscopically. Nucleic acid-based methods assess tumor-specific genetic alterations by analyzing whole cells or extracted RNA or DNA by PCR, RT-PCR, and whole-genome amplification. Due to interference caused by the expression of normal cell markers, nucleic acid based methods usually have a low specificity, but a high sensitivity [21,69].

Quantum dots (QDs) are an example of fluorescent NPs with size-dependent fluorescent emission that can be applied in the field of CTC detection. In comparison to fluorescent proteins and traditional dyes, QDs display a high quantum yield, tunable emission wavelengths and long fluorescence duration which can enhance the sensitivity of surface-marker dependent CTC capture [59]. It is also possible to capture heterogeneous CTCs by using simultaneous multicolor labeling of size-dependent QDs [59,70]. Due to their strong and stable fluorescence, QDs-based ex vivo CTC detection is highly relevant for clinical applications. However, the use of QDs for in vivo CTC detection provoked heavy metal toxicity [71].

3.3. Gold Nanoparticles

Gold nanoparticles (Au NPs) are another type of NP extensively used for improving the efficiency of CTCs enrichment and capture due to enhanced light absorption and scattering properties. In previous studies a variety of Au NPs have been synthesized exhibiting different shapes such like nanospheres, -rods or -shells. Subsequently, it is possible to functionalize the surface of these NPs with therapeutic agents, targeting moieties and imaging labels. The interaction of Au NPs and CTCs can be monitored by analyzing the protein adsorption processes at the Au NPs surfaces. By using surface plasmon resonance (SPR), the molecular adsorption is demonstrated by a measurable shift. Furthermore, it is possible to measure the binding between Au NPs and CTCs by using photoacoustic signals [22,58,72–74].

Based on unique characteristics, such as high sensitivity, flexibility and throughput, Au NPs are broadly applicable in the field of imaging and diagnostics. One example for an in vivo application is CTCs targeting by injection of Au NPs into the blood stream. This allows real-time, in situ monitoring of CTCs without blood sampling, sample preparation and the following CTCs isolation steps. Furthermore, this application enables the phagocytic clearance of CTCs upon binding [22,58]. Besides these advantages, the method also exhibits some disadvantages due to the particular conditions in the blood system, like high shear stress or immune response. Similar to other CTC targeting methods, injecting Au NPs into the blood may produce false positive results [75]. PEGylation of Au NPs is a commonly used strategy to overcome some of these issues resulting in an extended circulation time and decrease of non-specific binding [76,77].

Modified Au NPs with CTC-specific ligands can be used for direct binding and separation of CTCs from patient blood as an ex vivo approach [71]. This CTC detection method demonstrates two advantages: first, it protects patients from the potential toxicity of labeled NM for CTC-capturing and secondary, it enables cultivation and analysis of the isolated cells. Furthermore, it is possible to bind and enumerate CTCs label-free by immobilization of Au NPs on a nanostructured surface. For example, a thiolated ligand-exchange reaction with Au NPs on a herringbone chip (NP-^{HB}CTC-Chip) was used to isolate and release cancer cells from whole blood by Park et al. Antibody-coated NPs were chemically and directly assembled onto the ^{HB}CTC-Chip. This application has several advantages in comparison to antibodies coupled on flat silicon oxide surfaces: (1) increasing the available surface area to improve specific interaction of cancer cells with antibodies; (2) release of cancer cells from the surface by disrupting the metal-thiol interaction; (3) usage of released cancer cells for ex vivo cell culture and further molecular analysis; and (4) the optimization of this method for application in complex surface topographies without additional changes in the process by using chemically self-assembled monolayers [22,78].

3.4. Graphene and Carbon Nanotubes

Graphene is arranged in a two-dimensional layer of sp^2 hybridized carbon atoms ordered in a honeycomb network. Additionally, it is the basic structural block of other allotropes such as graphite and carbon nanotubes (CNTs). Unique chemical and physical properties of graphene and graphene oxide (GO) include strong mechanical strength, high surface area, high intrinsic mobility and great thermal conductivity with optical transmittance and electrical conductivity [58,79,80]. The chemical response results in a charge transfer between graphene and adsorbed molecules. GO can be functionalized through PEG-based chemistry and GO size is controllable by sonication time and filtration [81,82]. Moreover, graphene and GO have been used for electrical CTC detection due to the excellent electromagnetic detection of small biomolecules [83] and found their application in biological and medical research by using optical transparency for imaging [84]. The application of a GO chip for sensitive CTC capture was achieved by self-assembled GO nanosheets on a gold-patterned silicon surface via a positively charged intercalating agent and functionalization with PEG [85]. Yoon et al. spiked cells of different cancer cell lines into buffer or blood samples and flowed through a GO chip. Spiked cells were captured due to the usage of anti-EpCAM antibody for substrate functionalization by cross-linker and biotin-avidin linker chemistry. Blood samples from patients with breast, pancreatic and early lung cancer were cultivated on the gold-patterned surface with GO sheets with a capture efficiency of 2–23 CTCs/mL [83].

Furthermore, Wu et al. established an electrochemical protocol for the measurement of two tumor specific markers, such as anti-EpCAM and anti-GPC3, on a captured tumor cell surface by application of a GO film-modified glass carbon electrode [86]. This method allows the marker-dependent capture of tumor cells and enumeration of captured cells by square-wave voltammetry. It was also possible to use detected cells for fluorescent imaging [80,86]. The cultivation of captured CTCs opens the possibility for further applications and analysis [86].

The above-mentioned CNTs demonstrate remarkable electrical, mechanical and physico-chemical properties and are composed of graphitic hollow filaments of alterable lengths reaching up to several hundred micrometers. CNTs are known as two types: single-walled (SWCNTs) that comprise a single cylindrical sheet of graphene and double-/multi-walled (MWCNTs) that are composed of several concentric, coaxial, rolled up graphene sheets. The size of CNTs differs with a diameter typically ranging from 0.4 to 3 nm for SWCNT and from 2 to 200 nm for MWCNT [87]. CNTs are synthesized by chemical vapor deposition [87,88]. Due to electronic properties, the conductance of CNTs can be detected by electron current signals and is depending on chemical binding and mechanical deformations. Carbon nanofibers (CNFs) are also included to the group of CNTs. CNFs defend a less perfect graphene sheet arrangement featuring layers of graphene nanocones, the so called 'cups' and usually denote 'stacked-cup carbon nanotubes' [89].

First experiments on CTC detection in blood were published by Shao et al. [90] Binding of breast cancer cells to functionalized SWCNTs lead to a measurable decrease of conductivity. This assay contains a sensing area that is able to detect potential CTCs with low protein expression. This application presents the advantage of using samples without enrichment steps for direct cancer cell testing on the one hand, and on the other the challenge of a very small volume of analyzed patient blood (<10 μL) bearing the risk of missing CTCs. There is also the possibility of CTC counting difficulties because the signal is determined by a single cell reaching the space between the electrodes. Another example demonstrates the application of MWCNTs on a sensitive CNT-based biosensor for detection of CTCs from whole blood samples based on binding of anti-EpCAM antibodies to cancer cells and resulting in an increased electron transfer resistance. The detection of cells was demonstrated as an electrical response which was proportional to the concentration of cancer cells [59,91].

4. *Nano Meets Micro*—Micro-Mized Tools for Tumor Research

Beside the above described nano-sized systems for CTC detection and isolation, there are also techniques extending to the micro scale. Since these methods are widely used in combination with NMs in order to complement and advance nanomedical applications, we also discuss selected examples in the following sections.

4.1. Microfibricated Filters

Membrane microfilter devices are a suitable tool for separation of CTCs from whole blood samples by cell size exclusion [13,46,92,93]. Whereas CTCs can vary in their size and shape, the typical smaller dimensions of blood cells are 5–9 μm for erythrocytes, 10–15 μm for granulocytes, 7–18 μm for lymphocytes and 12–20 μm for monocytes [93]. The size exclusion approach is composed of a parylene-based membrane microfilter device including two parylene membrane layers and a photolithography-defined gap to minimize stress. This is the reason why isolated cells are viable and can be used for further molecular analysis [92,93]. The possibility of label-free isolation of CTCs is a large advantage of this technology. However the sensitivity to cell-size in a blood sample can lead to the risk of losing CTCs which are smaller than the filter pores because of their size and shape heterogeneity [46]. As an example, a parylene-based membrane microfilter device with integrated electrodes containing 11 μm diameter circular pores was used to isolate cancer cells. These cells were pre-stained with hematoxylin and spiked into a blood sample. This cell suspension was loaded into a syringe and dispensed to pass through the filter. The flow-through was collected by the bottom syringe. After the isolation process, immunostaining was used to determine potential CTCs from other cells on the filter. The recovery rate of the membrane filters was evaluated by hemocytometer using the hematoxylin staining of spiked cancer cells and resulted in 89.0 ± 9.5% recovery from blood [93].

4.2. Microbubbles in Diagnosis and CTC Detection

MBs are gas-filled, echogenic bubbles with a diameter typically comprised between 0.5 and 10 μm commonly used as contrast agents (CAs) in medical imaging and as carriers for targeted drug delivery, recently also gaining attention in the field of cancer diagnosis and treatment [94,95]. They consist of a low solubility complex gas, such as a perfluorocarbon (PFC) gas, surrounded by an external shell generally composed of phospholipids. A mixture of lipids in chloroform is homogenized by sonication in the presence of gas. PFC is especially suitable due to its low solubility in water, which is necessary for maintaining MB stability in the aqueous phase [96]. The bubble size is predominately determined by the solubility degree and partial pressure of the gas [96,97].

The MBs most investigated have soft shells made of phospholipids, sometimes of denaturated albumin, and contain a fluorocarbon (FC) as or among their inner gas phase component(s).

Standard shell phospholipids include dimyristoylphosphatidylcholine (DMPC) and dipalmitoyl-phosphatidylcholine (DPPC) [96,98]. The longer chain distearoylphosphatidylcholine (DSPC) forms semi-crystalline liquid-condensed monolayers that confer additional shell rigidity and stability. Pegylated lipids can provide stealthiness. The lipids and PEG chains can be fitted with a wide range of ligands. Polymeric shells can provide some additional stability but their response to UlS waves is usually dampened. The biologically inert inner FC gas, by considerably increasing lipid-coated bubble stability, made the development of commercial CAs a reality. The FC stabilizes MBs by drastically reducing the solubility of the inner gas in the continuous aqueous phase, by osmotically stabilizing the gas phase, and by providing co-surfactant activity with the phospholipids [96,99].

Furthermore, MBs were developed as CAs for conventional ultrasound (UlS)-imaging diagnostics and have been used for daily clinical practice for more than 20 years (Figure 4). Given their size and rheology comparable to red blood cells, MBs freely circulate through vessels and capillaries, with an average lifetime of 5 min. UlS triggers MBs resonance, resulting in detectable harmonic signals. Molecular imaging can be performed with MBs containing targeting ligands on the bubble-shell. During systemic circulation, targeted MBs progressively accumulate in the regions expressing the targeted molecules, defining areas of bright signals on UlS pictures [100,101]. As a consequence, the use of microbubbles for UlS-imaging can enhance their quality by precisely defining the region of targeted MBs accumulation. For example, MBs can attach to the vascular endothelium via a specific ligand-receptor bond, so that pathophysiological processes (like inflammation, angiogenesis, thrombosis, and tumors) can be imaged [101]. Multiple MBs fitted with targeting devices have been reported that are able to seek inflammation sites, myocardial ischemia, ischemia-reperfusion injury, ischemic memory, atherosclerotic plaques, thrombi, and angiogenesis in malignant solid tumors for the purpose of molecular imaging and focused therapy [102]. A perfluorobutane/lipid MB (BR55, Bracco, Milano, Italy) fitted with a heterodimeric peptide that has affinity for the vascular endothelial growth factor receptor type 2 (VEGFR2), a molecule expressed in neoangiogenesis, and hence, can help detect angiogenesis, has recently (2016) been licensed by the FDA for molecular imaging and characterization of liver masses and as intravesical contrast agent for voiding cystoureterography in children. This agent and other targeted MBs are also being investigated for detection of prostate, ovary and breast cancers [103–105].

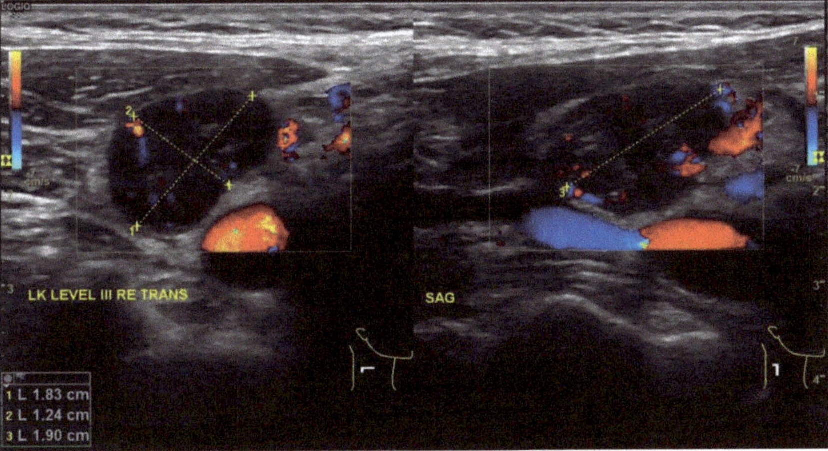

Figure 4. Ultrasound images of right transverse lymph node level III. Lymph node demonstrates malignant characteristics: axes are larger than 1 cm (left image), round shape with necrotic areas (right image).

Effective detection and isolation of CTC cells require high specificity and sensitivity, and simple and cost-effective technology. Targeted FC MBs are among the devices that are being investigated for CTC detection. They are easy to produce and collect cost-effectively. They can readily be fitted with multiple ligands, antibodies, and other markers that can recognize CTCs [102,103,106–109]. Due to their low density and buoyancy in aqueous media, they can easily be separated from aqueous media by flotation or gentle centrifugation. Once harvested, the isolated CTCs can be characterized, which can help determine treatment. CTCs may also be cultivated for personalized/precise drug testing. Besides their use for CTC isolation, MBs seem to enable CTC-targeted cancer drug delivery [102] (see Section 5.4) and may serve as carriers for drugs, genes and various markers, to overcome pathophysiological barriers, such as the blood brain barrier. Notably, MBs are furthermore already used for the delivery of energy, thus enabling techniques such as tissue ablation, sonothrombolysis or embolotherapy [102,110–112]. These approaches are technologically manageable, efficient and do not require any complex equipment. Their stability is limited, however, particular in biological fluids. But progress in formulation and preparation procedures has provided MBs sufficiently stable for efficient patient examination and for analytical procedures. MBs are extremely sensitive to UlS waves and can be imaged, monitored and manipulated by UlS [96,106]. Other imaging modalities, including ^{19}F-MRI can often be applied [113,114].

The use of MBs for CTC detection is very promising in theory but remains limited so far. Indeed, only one procedure using MBs to assist CTC separation from biological fluids has been reported [108,115]. This method is called flotation separation. The potential for targeted FC MBs to selectively bind and separate by buoyancy certain populations of circulating blood cells, initially erythrocytes and B-lymphoma cells, has been established [115]. This buoyancy-separation principle originally allowed for the isolation of CD4+ T lymphocytes from peripheral blood, following mixture with glass MBs [116]. In oncology, the capture of tumor cells has been demonstrated in solutions, blood or large-volume buffy coats [108,115]. Pancreatic tumor cells were captured within 15–30 min of incubation with functionalized MBs. The MB binding efficiency to human lung and mouse breast carcinoma into BSA/PBS or blood (around 90%) was comparable to that of commercial anti-EpCAM-coated magnetic beads (DynaBeads), ranging between 60–90% [117]. The EpCAM-targeted MBs can bind efficiently (85%) and rapidly (within 15 min) to various epithelial tumor cells suspended in cell medium [109]. In plasma-depleted blood, such MBs isolated tumor cells at high (105–106 cells/mL) and low (10–20 cells/mL) concentrations of tumor cells (mouse breast 4T1, human prostate PC-3 and pancreatic cancer BxPC-3 cells). However, in whole blood, MBs presented decreased stability, possibly due to gas mixing and exchange. Further development of the method led then to the design of blood-stable MBs for isolation of breast tumor cells [102]. Parallel studies on the lipid shell functionalization with anti-human EpCAM or EGFR antibodies demonstrated different preferential binding abilities to several breast tumor cell lines with distinct marker expression profiles, and culminated in the production of multi-targeted anti-human EpCAM/EGFR MBs recognizing all cell lines with over 95% efficiency [102]. Fast (30 min) and efficient (70–90%) recovery of CTCs was achieved in human blood. In patients with metastatic breast cancer, these MBs allowed for the isolation of CTCs, cell clusters and tumor-derived CK+/CD45- microparticles. Also, albumin-based MBs have been used for buoyancy-activated cell sorting, which showed inherent advantages (such as stability and simplicity of formulation) with respect to lipid-based ones [118]. In albumin-systems, the most common way for antibody conjugation is based on the non-covalent incorporation into the albumin shell of avidin linkers as anchor sites for biotinylated antibodies. Nevertheless, to strengthen the antibody conjugation, biotin can be first connected by a covalent amide bond to albumin, followed by incubation with avidin and biotinylated antibodies. These biotin-MBs targeted against CD44 receptors efficiently recognized luminal breast cancer cells in PBS, and were able to separate them from the CD44- basal-like breast cancer subset with higher sorting purity than other control MBs [118]. These results contribute to establishing that targeted MBs are an effective new approach to liquid biopsy [119]. As compared to other methods (adherence, absorbance, particle size, density gradient, dielectric properties, chemo-resistance), the antigen-antibody recognition provides precise sorting. The two major sorting tools, being fluorescence activated cell sorting (FACS) and MACS, use expensive and large instruments, long processing time, and magnetic forces that may damage some types of cells [120–122]. Similarly, microfluidic approaches exert substantial shear stresses thus risking cell damage [123]. Instead, buoyant MBs isolate specific cells with molecular precision in a simple and safe way, given that the shear stress originated from a rising bubble and the tension from the buoyancy force are both far below the threshold for cell damage [124]. Moreover, whereas the assays where nano- or micron-sized immunomagnetic beads capture the CTCs suffer from some other limitations (such as non-specific carryover, relatively long processing time and contamination with leukocytes) [125–127], the MB-assisted cell isolation emerges as a promising method for rapid and accurate collection of exfoliated tumor cells in a variety of pathological samples (e.g., blood, bone marrow, urine). Finally, besides being cost-efficient and scalable, the flotation separation technology presents also wide horizons of optimization for biological use, considering the large versatility of MB surface functionalization.

MBs can also deliver therapeutic agents, and hence, exhibit extended theranostic potential. Remarkably, it was found that exposure to a supernatant FC gas can significantly enhance the adsorption and retention of a large variety of molecules, including lipids, proteins, surfactants, poloxamers, fluorinated drugs and hypoxia biomarkers at the surface of lipid-shelled MBs. The same phenomenon has been observed with diverse NPs, such as magnetic iron oxide, cerium oxide or nanodiamonds. The effect is particularly marked for fluorinated molecules and particles.

Multiple hybrid MBs are also being developed that carry NPs enclosed in or attached to their shell, including superparamagnetic iron oxides [128,129], QDs [130], gold clusters or nanorods [102,131,132], GO sheets [133], cerium oxide NPs [134], liposomes [135]. Small and stable MBs decorated with dendronized iron oxide magnetic NPs were obtained that are stabilized by fluorine-fluorine interactions between the internal FC gas and the fluorinated terminal end-group of the oligo(ethylene glycol)-based dendrons [136].

4.3. Microfluidic Lab-on-a-Chip Devices

A new development in the field of CTC enrichment and detection includes microfluidic lab-on-a-chip devices with immense advantages including cost-effectiveness, miniaturization, and the improvement of efficiency since it could be integrated into other techniques [14,137]. Consequently, isolation and analysis of CTCs on one chip can improve the number of catched CTCs by avoiding loss of rare cells during the experimental steps of sorting, enumeration and analysis [138]. Current microchip platforms are based on magnetic force, affinity, size or other physical properties and are separated into two types of microfluidic devices for CTC detection [139]. The first type of microfluidic devices includes the immunomagnetic-based method for CTC detection (e.g., CTC chip) and the second type represents the method of antibody-labeling combined with physical isolation (Figure 5), which can consist of different materials like silicon, glass or thermopolymer.

4.4. Immunomagnetic-Based Method and 'Micro-Hall Detector'

Immunomagnetic-based CTC chip separation is performed by using the advantages of two combined techniques: the immunomagnetic separation and the microfluidic device. The capture efficiency depends on the magnetic strength and drag force under the flow condition. Cells bound to large number of NPs can be captured more efficiently by using both forces [21]. The isolation of cells in microfluidic channels is performed in the presence of a permanent magnetic field which can be located under the bottom of the chip [80,140] or on top of the channel to improve the separation efficiency by inverting the microchannel that results in gravity direction opposite to the magnetic field [141].

The possibility for a fast screening method for CTCs in a blood sample with a miniaturized microfluidic technology is called micro-Hall detector (μHD). The μHD can selectively and sensitively detect a wide range of single cellular biomarkers or multiple biomarkers on individual cells as screening system (Figure 5A). MNPs-immunolabeled cells can be detected via monitoring the magnetic moments of cells in-flow on a single microfluidic chip. There is also an option to use MNPs with different magnetization properties to label different cellular markers. By using the particles' classifiable magnetization properties, the quantity of each MNP type representing the expression level of a distinct target biomarker in a single cell can be obtained [142].

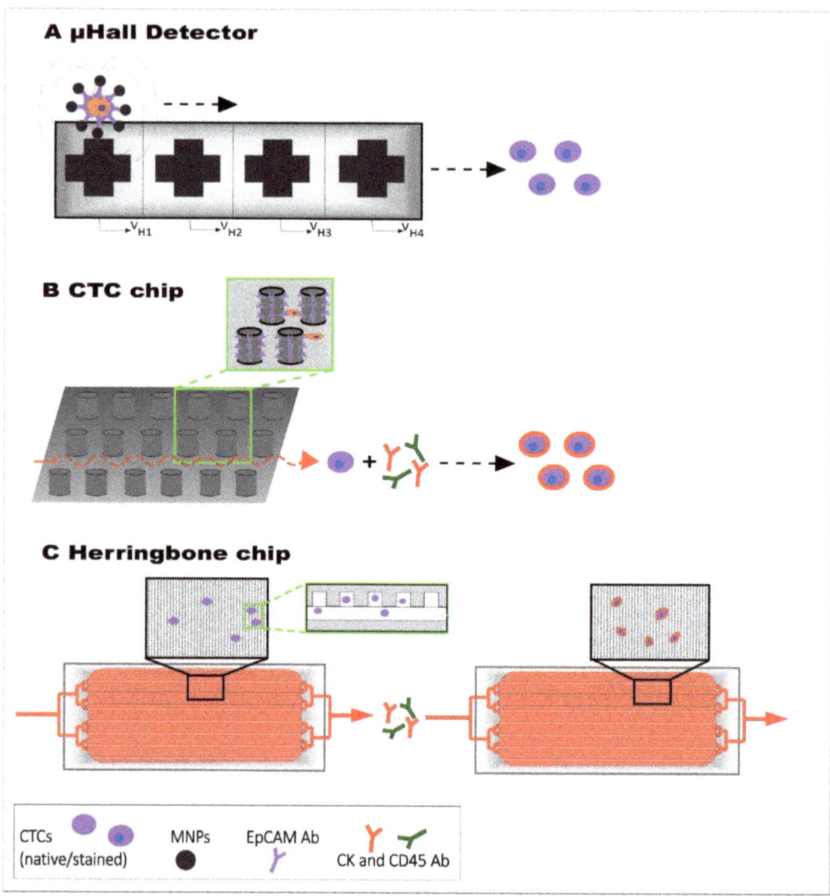

Figure 5. Design of microfluidic chips for CTC detection. Whole blood sample is pushed through the surface of the chip. (**A**) Cells are MNPs-immunolabeled and can be detected via monitoring the magnetic moments of cells in-flow on μHall detection chip. (**B**) CTC chip is coated with a CTC-specific antibody, such as EpCAM, and contains Ab-coated microposts. This system is also used in herringbone chip that contains Ab-coated microchannels (**C**). Captured cells are stained for CK, CD45 and DAPI for identification and enumeration.

4.5. 'CTC –Chip' as Silicon-Based Alternative

Another microfluidic device used for efficient and reproducible isolation of CTCs from the blood of patients with common epithelial tumors is called 'CTC-chip' [143]. This microfluidic system is composed of three parts: the CTC-chip etched in silicon, a manifold to enclose the chip, and a pump producing the flow through the capture module (Figure 5B). Additionally, the CTC-chip contains an EpCAM-antibody functionalized array of microposts. The cell capture efficiency can be influenced by two essential parameters: flow speed and shear force. The flow speed is important due to its influence on the duration of cell-micropost contact, whereas the shear force has to be minimized to guarantee a high cell-micropost attachment [143].

4.6. Glass/PDMS-Based Chip and 'Herringsbone-Chip'

After the development of the CTC-chip a modified herringbone CTC capture chip was developed to increase the interaction of flowing cells and anti-EpCAM-functionalized polydimethylsiloxane (PDMS) microchannels through passive mixing [144] (Figure 5C). The 'herringbone-chip' has integrated microvortices to disrupt streamlines and increase the capture efficiency. Due to the antibody-antigen interaction, cells tether to the chip and can be stained afterwards. The advantage of this glass chips is the transparency that allows clear imaging by using different types of light microscopy based techniques [80].

4.7. Thermoresponsive Polymer-Based CTC Chip

Due to magnificent optical transparency and low costs polymethylmethacrylate (PMMA) is also used for microfluidic CTC capture and analysis. PMMA includes UV exposure generated carboxylic acid groups on the surface to analyze protein concentration, electroless deposition and cancer cell capture [145]. Due to the enhanced surface area for functionalization, the surface area roughness can be additionally increased by high intensity light. The thermal bonding passes through to preserve these microfeatures at low temperature [146]. Accordingly, for CTCs specific capture and enumeration a high-throughput microsampling unit functionalized with anti-EpCAM antibodies and an included conductivity sensor has been developed [80].

5. Applications in Nanomedicine

As fighting tumor metastasis is, besides elimination of the primary tumor, the overarching goal of chemotherapy, therapeutic targeting and specific depletion or destruction of CTCs from blood vessels may be an intriguing strategy for prevention of tumor metastasis. With the emerging possibilities of nanoscale materials, researchers have new tools at hand to design and develop a variety of nanosystems for targeted delivery of therapeutic agents to CTCs hoping to efficiently destroy them and thereby to inhibit tumor metastasis (Figure 6).

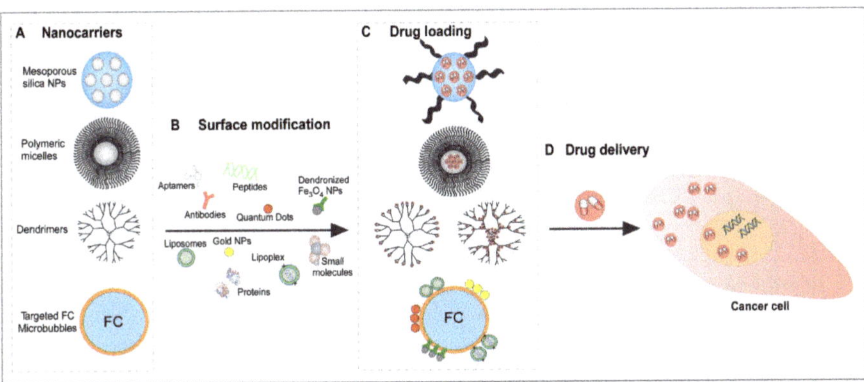

Figure 6. Illustration of drug delivery system of different nanocarriers: (**A**) Unloaded nanocarriers: Mesoporous silica NPs, polymeric micelles, dendrimers and targeted FC microbubbles. (**B**) Surface modification of nanocarriers with cancer cell specific targets (aptamers, antibodies, dendronized Fe_3O_4 NPs, gold NPs, liposomes, lipoplex, peptides, proteins, quantum dots, and small molecules). (**C**) Illustration of drug loaded nanocarriers and drug release in cancer cells (**D**).

5.1. Mesoporous Silica Nanoparticles

Mesoporous silica nanoparticles (MSNs) with a pore size from 2 nm to 50 nm show attractive properties, such as a large surface area, mesoporous structure as well as a controllable pore size, ease of surface functionalization and good biocompatibility [147] and have therefore drawn considerable attention especially for drug delivery applications during the past years. For example, nanoplatforms have been developed, which can specifically target colorectal cancer cells via an EpCAM antibody-functionalization representing an in vitro model for CTC targeting in colorectal cancer patients [148]. Furthermore, loading of MSNs with mifepristone allowed the inhibition of lung metastasis in mice. Thus, this proof-of-concept study suggested the MSN-based prevention of metastasis spread by restraining CTC activity [148].

5.2. Polymeric Colloidal Particles as Polymeric Micelles

Polymeric colloidal particles can be produced in a size range of several nm and allow for secure encapsulation, adsorption or conjugation of therapeutic agents within their polymeric matrix or their surface [149,150]. Due to their biodegradability and biocompatibility as well as the possibility to combine a broad range of controlled chemical and physical properties by molecular synthesis, polymeric materials remain an important cancer drug delivery system. Linear, globular or branched polymers of different sizes have been used in the past [151–153]. Core-shell particles, which are mainly formed using amphiphilic block copolymers, are called "micelles". Micelles consist of a hydrophobic core to minimize aqueous exposure and a hydrophilic shell to stabilize the core [154]. These properties leave micelle structures attractive for drug delivery applications as they allow the loading with hydrophobic small molecule drugs into their core while a steric protection can be added to the outer "shell" layer. Additionally, hydrophilic drugs including macromolecules like nucleic acids can be included into polymeric NPs by electrostatic attraction or chemical conjugation. Moreover, controlled release of macromolecules from micelles has now been applied in different studies [155,156].

In the design of different drug-loaded nano-systems, a number of materials have been applied and were found to be well suited, albeit showing different advantages depending on the application. Polyamides, poly(amino acids), polyesters and polyacrylamides with thermoplastic aliphatic polyesters such as poly glycolic acid, polylactic acid (PLA) and copolymer poly (lactic co-glycolic acid) (PLGA) are some of the most common examples [157–160]. Due to its high biodegradability, PLGA is often utilized in biomedical applications [156,161]. Furthermore, PLA and chitosan form polymeric micelles [149] whereas the latter can be applied as transport vehicle for hydrophilic drugs. Similarly, PLGA and PLA exhibit advantageous characteristics such as low toxicity in combination with negative surface charge [162]. Nonspecific side effects of the antitumor agent Doxorubicin (Dox) could be reduced by Dox encapsulation in chitosan NPs tested for the treatment of solid tumors in vivo [163,164].

Deng et al. usd Dox-loaded biodegradable polymeric micelles to target CTCs and finally suppress tumor metastasis [165]. They synthesize monomethyl poly (ethylene glycol)-poly (e-caprolactone) (mPEG-PCL) deblock copolymers to prepare Dox-loaded micelles via a pH-induced self-assembly. Whereas unloaded micelles showed minimal cytotoxicity when incubated with 4T1 cells even at very high concentrations, micelles loaded with Dox induced slightly higher cytotoxicity than Dox alone. Dox micelles were able to inhibit tumor growth, suppress tumor metastasis by killing CTCs and extend the survival rate in transgenic zebrafish as well as a mouse model, by inducing apoptosis and reducing the number of proliferation-positive cells in tumors [165].

In another study, a designed nanoplatform consisting of paclitaxel-loaded PEG-PLA polymeric micelles was successfully applied to achieve dual damaging of the primary tumor as well as CTCs [166]. In a recent study, Gener et al. loaded polymeric micelles with the FDA-approved drug Zileuton™ which has been reported to be a potent inhibitor of cancer stem cells. Interestingly, the authors reported complete eradication of CTCs in the blood stream of an in vivo mouse model and thus, suggested their system to be effective against metastatic spread [167].

5.3. Dendrimers

Dendrimers are highly soluble, can be synthesized with a uniformity in size and composition, and have a high number of surface groups. This combination of properties and the ease of functionalizing the surface groups, make them interesting to develop drug development strategies [168]. Employing a novel polyamidoamine dendrimer-based nanoplatform, CTCs could be captured captured and their adhesion to the vascular endothelial layer inhibited [169–172]. The nanoscale dendrimers therefore made use of two antibodies coated to their surface and targeting membrane markers of human colorectal CTCs (anti-EpCAM and -Slex). Whereas it has been reported, that targeting of EpCAM can directly disturb the adhesion process of CTCs, the Slex (saliva acidifying louis oligosaccharides X) antibody can indirectly interrupt the adhesion between CTCs and endothelial cells via Slex/E-selection interaction [169]. In comparison to their single antibody-coated counterparts, the dual antibody conjugates displayed a remarkably enhanced efficiency and specificity in recognizing and capturing CTCs from a large population of leukocytes or red blood cells *in vitro*, as well as from the blood of patients and mice *in vivo*. Recently, this group developed dual aptamer rings which are conjugated on dendrimers and thus, are able to simultaneously target EpCAM and Her2 biomarkers on CTCs in the presence of millions of normal cells with excellent stability and accuracy [173]. The described study provides new ideas for the design of more powerful and intelligent nanomedicines allowing the prevention of tumor metastasis via suppressing CTCs and blocking their adhesion to blood vessels. Zheng et al. presented a type of barcode particle consisted of spherical colloidal crystal clusters which are surrounded by dendrimer-amplified aptamer probes [174]. A specific aptamer functionalization let the particles interact with specific CTC types and used dendrimers able to amplify the effect of the aptamers. Particles with these capabilities are able to capture, detect and release multiple types of CTCs from clinical samples [174].

5.4. Microbubbles in Tumor Treatment

In recent years, MBs have started to be used for therapeutic approaches [175]. Chemotherapeutic agents can be delivered to malignant tissues by combining UlS and MBs, enhancing the in vivo delivery of the drug into the tumor, thereby minimizing the harmful systemic side effects on normal tissues [176]. In this regard, focused UlS in the presence of circulating MBs have been extensively exploited to temporarily open the blood-brain barrier and elicit the passage of chemotherapeutic agents into neural neoplasms. In fact, the local tumor insonation elicits a stable *cavitation* effect, as the UlS energy is transferred to the circulating MBs, letting them to expand and contract cyclically. This oscillation results in damage of the tight junctions and interruption of the contiguous endothelial cell layer [177].

Novel MBs have been developed to directly be loaded with drugs and to release them, thus acting as drug delivery platforms, either in the presence or in the absence of UlS stimulation. Strategies for incorporation of several drugs onto the MB surface have been reviewed [107,178]. The mechanisms of drug release and tumor delivery are different. For instance, the cavitation can cause MB rupture during tumor insonation and drive the drug throughout the capillary wall by delivering a *"ballistic effect"* [179]; alternatively, sonication can induce MB oscillation leading to permeabilization of the contiguous cell membranes, enhancing the entry of a locally released agent into either cancerous or endothelial cells [180,181]. Furthermore, MBs loaded with specific molecules, called *sonosensitizers*, can be used for sonodynamic tumor therapy [175,182]. In sonodynamic therapy, cell cultures or tumors are sonicated by UlS at selected frequency and intensity range (usually around 1.0–2.0 MHz, 0.5–3.0 W.cm^{-2}), resulting in phenomena of inertial cavitation. The rapid collapse of bubbles in the liquid milieu determines shock waves producing free radicals and a cascade of molecular events that activate the sonosensitizers, which in turn kill rapidly dividing cancer cells nearby [183]. Conventional sonosensitizers are the same light-sensitive agents developed for photodynamic therapy [184], like hematoporphyrin and its derivatives. When loaded onto MBs [175], their circulation into the tumor vasculature can then be monitored by diagnostic UlS, so that the sonodynamic process can be initiated once detected within the mass, thus creating a therapeutic-diagnostic platform to monitor the treatment effectiveness [185–187]. In addition, damaging of the tumor vascularity can also be achieved through MBs (i.e., *antivascular therapy*). Here, MBs act as vascular disrupting agents, whose insonation results into local thermal and cavitation effects leading to the destruction of endothelial cells lining the tumor vessels [188] and necrosis of the neoplastic cells, with a consequent reduction in tumor growth and lengthened survival time. Finally, targeted MBs could be used to selectively eliminate CTCs during their systemic circulation: after intrasystemic injection and CTCs recognition, their insonation might induce local cavitation effects and stimulate either release of chemotherapeutics or cell damage. Moreover, the possibility to combine MBs with other nano-objects (such as liposomes or superparamagnetic NPs) widens the plethora of therapeutic options that one could exploit for CTC killing, including new strategies of drug loading/release and thermotherapy. Nevertheless, even though the principle of the CTC-eliminating MBs could be easily applied, thus far only one study of such kind has been carried out [189]: here, liposome-loaded MBs targeted to N-cadherin (N-cad) could bind to a human melanoma cell line derived from a lymph node metastasis (HMB2 cells). Upon insonation, a model drug (propidium iodide) loaded onto the liposomes was intracellularly delivered to N-cad-expressing cells only. Due to the great potential of MB-mediated CTC depletion strategy and to the urgency in finding effective and portable solutions to hinder cancer relapse caused by metastatic colonization, more efforts in development of such technology are expected in a near future.

6. Conclusions and Outlook

In the field of nanomedicine, the application of engineered NMs has assumed an increasing role in early cancer diagnosis and efficient treatment. The analysis of captured CTCs in liquid biopsies from cancer patients provides important information about the biology of cancer micrometastases, and offers a well-tolerated alternative to standard biopsies in the clinical management of carcinoma patients.

Preliminary results of CTCs' enumeration and analysis obtained by the FDA-approved CellSearch system suggest the possibility of 'on-line' monitoring of an ongoing therapy and the drug efficiency. In recent years, a variety of CTC isolation assays have been evolved for supervising a range of distinct tumor types at different disease stages. Due to the extremely rare presence of CTCs in peripheral blood the isolation and detection of CTCs can be very challenging. Consequently, new technologies have to accomplish the challenges of rare cells physical properties including their size, density, deformability and cell shape. Therefore, specificity and sensitivity and cost-effectiveness remain the key issues which upcoming technologies need to address. The development of nanotechnology-based methods for detection and follow-up analysis of CTCs represents a milestone to achieve high capture efficiency, accuracy, and sensitivity. NMs can implement the possibility of a multiplexed targeting because of their possibility to be modified with different targeting ligands to capture, isolate and detect CTC subpopulations.

By summarizing a variety of NM-based enrichment, capture and also detection methods, and by comparing them to complementary micro-sized systems, such as the use of microbubbles, the advantages, but also some disadvantages of nano-sized systems became obvious. Obvious advantages of nanostructured substrates and platforms for the detection and capture of CTCs are better ligand-antigen binding between the functionalized substrate and captured CTCs. Nanostructured substrates demonstrate enhanced local topographic interactions that lead to enhanced cell capture affinity. Additionally, ligand coating of nanostructures can be prepared with much higher density thereby improving binding affinity in comparison to micro- and macrostructures. Moreover, the use of microfluidic chips as cost-effective, miniaturized and efficiency improved applications for the enrichment and detection of CTCs obtain better performance with nanostructured substrates. In contrast, few nanomaterials have made it to clinical trials, or even clinical practice and there is not yet any FDA-approved nanomedical product that markedly improves patient survival or quality of life. The use of NMs in nanomedicine products (e.g., devices and therapeutics) has been quite limited due to their frequent toxic and harmful properties in biological and medical contexts [190]. To however improve therapeutic gain of nanomedicine also in the CTC field, a mechanistic understanding of determinants at nanobio-interfaces is a must.

The benefits of microchip technology and nanotechnology are associated with a combination of NMs with microfluidic devices to optimize the CTC capture methods for further analysis. Targeted lipid-coated FC MBs may ensure rapid and efficient isolation by simple flotation of CTCs from the blood of patients with metastatic cancer, and enable focused drug delivery as well. Besides detection and analysis of CTCs, a huge potential of this technology lies in a CTC-targeted cancer therapy to eliminate tumor cells in the peripheral blood. In conclusion, the further development of nano-sized CTC systems should be not only focused on tumor diagnosis and monitoring, but should also exploit its own potential as anticancer treatment. To benefit from a combination of available technologies, the "nano meets micro" approach seems to be promising to achieve long overdue progress in the field of nanomedicine.

Author Contributions: All authors contributed toward data analysis, drafting and critically revising the paper, gave final approval of the version to be published, and agree to be accountable for all aspects of the work. All authors have read and agreed to the published version of the manuscript.

Funding: This work was supported by grants from the Research Center for Natural and Medical Sciences ([Naturwissenschaftlich-medizinisches Forschungszentrum]; NMFZ), Brigitte and Konstanze Wegener Foundation, German Research Foundation ([Deutsche Forschungsgemeinschaft]; DFG), foreign experts project of Shanxi Provincial '100 Talents Plan' (RS, GD), and NanoTransMed, which is co-funded by the European Regional Development Fund (ERDF) in the framework of the Interreg V Upper Rhine program, the Swiss Confederation, and the Swiss cantons of Aargau, Basel-Landschaft, and Basel-Stadt.

Conflicts of Interest: The authors declare no conflict of interest.

References

1. Docter, D.; Westmeier, D.; Markiewicz, M.; Stolte, S.; Knauer, S.K.; Stauber, R.H. The nanoparticle biomolecule corona: Lessons learned—Challenge accepted? *Chem. Soc. Rev.* **2015**, *44*, 6094–6121. [CrossRef] [PubMed]
2. Mocellin, S.; Hoon, D.; Ambrosi, A.; Nitti, D.; Rossi, C.R. The prognostic value of circulating tumor cells in patients with melanoma: A systematic review and meta-analysis. *Clin. Cancer Res.* **2006**, *12*, 4605–4613. [CrossRef] [PubMed]
3. Pantel, K.; Alix-Panabieres, C. Circulating tumour cells in cancer patients: Challenges and perspectives. *Trends Mol. Med.* **2010**, *16*, 398–406. [CrossRef] [PubMed]
4. Kramer, O.H.; Stauber, R.H.; Bug, G.; Hartkamp, J.; Knauer, S.K. SIAH proteins: Critical roles in leukemogenesis. *Leukemia* **2013**, *27*, 792–802. [CrossRef] [PubMed]
5. Garzia, L.; D'Angelo, A.; Amoresano, A.; Knauer, S.K.; Cirulli, C.; Campanella, C.; Stauber, R.H.; Steegborn, C.; Iolascon, A.; Zollo, M. Phosphorylation of nm23-H1 by CKI induces its complex formation with h-prune and promotes cell motility. *Oncogene* **2008**, *27*, 1853–1864. [CrossRef] [PubMed]
6. Hong, Y.; Fang, F.; Zhang, Q. Circulating tumor cell clusters: What we know and what we expect (Review). *Int. J. Oncol.* **2016**, *49*, 2206–2216. [CrossRef]
7. Lambert, A.W.; Pattabiraman, D.R.; Weinberg, R.A. Emerging Biological Principles of Metastasis. *Cell* **2017**, *168*, 670–691. [CrossRef]
8. Nguyen, D.X.; Bos, P.D.; Massague, J. Metastasis: From dissemination to organ-specific colonization. *Nat. Rev. Cancer* **2009**, *9*, 274–284. [CrossRef]
9. Austin, R.G.; Huang, T.J.; Wu, M.; Armstrong, A.J.; Zhang, T. Clinical utility of non-EpCAM based circulating tumor cell assays. *Adv. Drug Deliv. Rev.* **2018**. [CrossRef]
10. Gribko, A.; Kunzel, J.; Wunsch, D.; Lu, Q.; Nagel, S.M.; Knauer, S.K.; Stauber, R.H.; Ding, G.B. Is small smarter? Nanomaterial-based detection and elimination of circulating tumor cells: Current knowledge and perspectives. *Int. J. Nanomedicine* **2019**, *14*, 4187–4209. [CrossRef]
11. Allard, W.J.; Matera, J.; Miller, M.C.; Repollet, M.; Connelly, M.C.; Rao, C.; Tibbe, A.G.; Uhr, J.W.; Terstappen, L.W. Tumor cells circulate in the peripheral blood of all major carcinomas but not in healthy subjects or patients with nonmalignant diseases. *Clin. Cancer Res.* **2004**, *10*, 6897–6904. [CrossRef] [PubMed]

12. Kim, S.; Han, S.I.; Park, M.J.; Jeon, C.W.; Joo, Y.D.; Choi, I.H.; Han, K.H. Circulating tumor cell microseparator based on lateral magnetophoresis and immunomagnetic nanobeads. *Anal. Chem.* **2013**, *85*, 2779–2786. [CrossRef] [PubMed]
13. Pantel, K.; Brakenhoff, R.H.; Brandt, B. Detection, clinical relevance and specific biological properties of disseminating tumour cells. *Nat. Rev. Cancer* **2008**, *8*, 329–340. [CrossRef] [PubMed]
14. Hao, S.J.; Wan, Y.; Xia, Y.Q.; Zou, X.; Zheng, S.Y. Size-based separation methods of circulating tumor cells. *Adv. Drug Deliv. Rev.* **2018**. [CrossRef] [PubMed]
15. Ried, K.; Eng, P.; Sali, A. Screening for Circulating Tumour Cells Allows Early Detection of Cancer and Monitoring of Treatment Effectiveness: An Observational Study. *Asian Pac. J. Cancer Prev.* **2017**, *18*, 2275–2285. [CrossRef]
16. Jaeger, B.A.; Jueckstock, J.; Andergassen, U.; Salmen, J.; Schochter, F.; Fink, V.; Alunni-Fabbroni, M.; Rezai, M.; Beck, T.; Beckmann, M.W.; et al. Evaluation of two different analytical methods for circulating tumor cell detection in peripheral blood of patients with primary breast cancer. *Biomed. Res. Int.* **2014**, *2014*, 491459. [CrossRef]
17. Murray, N.P.; Albarran, V.; Perez, G.; Villalon, R.; Ruiz, A. Secondary Circulating Tumor Cells (CTCs) but not Primary CTCs are Associated with the Clinico-Pathological Parameters in Chilean Patients With Colo-Rectal Cancer. *Asian Pac. J. Cancer Prev.* **2015**, *16*, 4745–4749. [CrossRef]
18. Wendel, M.; Bazhenova, L.; Boshuizen, R.; Kolatkar, A.; Honnatti, M.; Cho, E.H.; Marrinucci, D.; Sandhu, A.; Perricone, A.; Thistlethwaite, P.; et al. Fluid biopsy for circulating tumor cell identification in patients with early-and late-stage non-small cell lung cancer: A glimpse into lung cancer biology. *Phys. Biol.* **2012**, *9*, 016005. [CrossRef]
19. Engel, H.; Kleespies, C.; Friedrich, J.; Breidenbach, M.; Kallenborn, A.; Schondorf, T.; Kolhagen, H.; Mallmann, P. Detection of circulating tumour cells in patients with breast or ovarian cancer by molecular cytogenetics. *Br. J. Cancer* **1999**, *81*, 1165–1173. [CrossRef]
20. de Bono, J.S.; Scher, H.I.; Montgomery, R.B.; Parker, C.; Miller, M.C.; Tissing, H.; Doyle, G.V.; Terstappen, L.W.; Pienta, K.J.; Raghavan, D. Circulating tumor cells predict survival benefit from treatment in metastatic castration-resistant prostate cancer. *Clin. Cancer Res.* **2008**, *14*, 6302–6309. [CrossRef]
21. Bhana, S.; Wang, Y.; Huang, X. Nanotechnology for enrichment and detection of circulating tumor cells. *Nanomedicine (Lond)* **2015**, *10*, 1973–1990. [CrossRef] [PubMed]
22. Zhang, Z.; King, M.R. Nanomaterials for the Capture and Therapeutic Targeting of Circulating Tumor Cells. *Cell. Mol. Bioeng.* **2017**, *10*, 275–294. [CrossRef] [PubMed]
23. Satelli, A.; Mitra, A.; Brownlee, Z.; Xia, X.; Bellister, S.; Overman, M.J.; Kopetz, S.; Ellis, L.M.; Meng, Q.H.; Li, S. Epithelial-mesenchymal transitioned circulating tumor cells capture for detecting tumor progression. *Clin. Cancer Res.* **2015**, *21*, 899–906. [CrossRef] [PubMed]
24. Lindner, J.R. Microbubbles in medical imaging: Current applications and future directions. *Nat. Rev. Drug Discov.* **2004**, *3*, 527–532. [CrossRef] [PubMed]
25. Rauscher, H.; Sokull-Kluttgen, B.; Stamm, H. The European Commission's recommendation on the definition of nanomaterial makes an impact. *Nanotoxicology* **2013**, *7*, 1195–1197. [CrossRef]
26. Ashworth, T.R. A Case of Cancer in Which Cells Similar to Those in the Tumors Were Seen in the Blood after Death. *Australas. Med. J.* **1869**, *14*, 146–149.
27. Arya, S.K.; Lim, B.; Rahman, A.R. Enrichment, detection and clinical significance of circulating tumor cells. *Lab Chip* **2013**, *13*, 1995–2027. [CrossRef]
28. Ming, Y.; Li, Y.; Xing, H.; Luo, M.; Li, Z.; Chen, J.; Mo, J.; Shi, S. Circulating Tumor Cells: From Theory to Nanotechnology-Based Detection. *Front Pharmacol.* **2017**, *8*, 35. [CrossRef]
29. Clevers, H. The cancer stem cell: Premises, promises and challenges. *Nat. Med.* **2011**, *17*, 313–319. [CrossRef]
30. Aceto, N.; Bardia, A.; Miyamoto, D.T.; Donaldson, M.C.; Wittner, B.S.; Spencer, J.A.; Yu, M.; Pely, A.; Engstrom, A.; Zhu, H.; et al. Circulating tumor cell clusters are oligoclonal precursors of breast cancer metastasis. *Cell* **2014**, *158*, 1110–1122. [CrossRef]
31. Wollenberg, B. Implication of stem cells in the biology and therapy of head and neck cancer. *GMS Curr. Top Otorhinolaryngol. Head Neck Surg.* **2011**, *10*. Doc01. [CrossRef] [PubMed]
32. Krawczyk, N.; Meier-Stiegen, F.; Banys, M.; Neubauer, H.; Ruckhaeberle, E.; Fehm, T. Expression of stem cell and epithelial-mesenchymal transition markers in circulating tumor cells of breast cancer patients. *Biomed. Res. Int.* **2014**, *2014*, 415721. [CrossRef] [PubMed]

33. Massague, J.; Obenauf, A.C. Metastatic colonization by circulating tumour cells. *Nature* **2016**, *529*, 298–306. [CrossRef] [PubMed]
34. Nieto, M.A. Epithelial plasticity: A common theme in embryonic and cancer cells. *Science* **2013**, *342*, 1234850. [CrossRef] [PubMed]
35. Banyard, J.; Chung, I.; Wilson, A.M.; Vetter, G.; Le Bechec, A.; Bielenberg, D.R.; Zetter, B.R. Regulation of epithelial plasticity by miR-424 and miR-200 in a new prostate cancer metastasis model. *Sci. Rep.* **2013**, *3*, 3151. [CrossRef] [PubMed]
36. Tsai, J.H.; Donaher, J.L.; Murphy, D.A.; Chau, S.; Yang, J. Spatiotemporal regulation of epithelial-mesenchymal transition is essential for squamous cell carcinoma metastasis. *Cancer Cell* **2012**, *22*, 725–736. [CrossRef] [PubMed]
37. Ocana, O.H.; Corcoles, R.; Fabra, A.; Moreno-Bueno, G.; Acloque, H.; Vega, S.; Barrallo-Gimeno, A.; Cano, A.; Nieto, M.A. Metastatic colonization requires the repression of the epithelial-mesenchymal transition inducer Prrx1. *Cancer Cell* **2012**, *22*, 709–724. [CrossRef]
38. Jie, X.X.; Zhang, X.Y.; Xu, C.J. Epithelial-to-mesenchymal transition, circulating tumor cells and cancer metastasis: Mechanisms and clinical applications. *Oncotarget* **2017**, *8*, 81558–81571. [CrossRef]
39. Mani, S.A.; Guo, W.; Liao, M.J.; Eaton, E.N.; Ayyanan, A.; Zhou, A.Y.; Brooks, M.; Reinhard, F.; Zhang, C.C.; Shipitsin, M.; et al. The epithelial-mesenchymal transition generates cells with properties of stem cells. *Cell* **2008**, *133*, 704–715. [CrossRef]
40. Lippert, B.M.; Knauer, S.K.; Fetz, V.; Mann, W.; Stauber, R.H. Dynamic survivin in head and neck cancer: Molecular mechanism and therapeutic potential. *Int. J. Cancer* **2007**, *121*, 1169–1174. [CrossRef]
41. Shen, Z. Cancer biomarkers and targeted therapies. *Cell Biosci.* **2013**, *3*, 6. [CrossRef] [PubMed]
42. Knauer, S.K.; Stauber, R.H. Development of an autofluorescent translocation biosensor system to investigate protein-protein interactions in living cells. *Anal. Chem.* **2005**, *77*, 4815–4820. [CrossRef] [PubMed]
43. Alix-Panabieres, C.; Pantel, K. Challenges in circulating tumour cell research. *Nat. Rev. Cancer* **2014**, *14*, 623–631. [CrossRef] [PubMed]
44. Opoku-Damoah, Y.; Assanhou, A.G.; Sooro, M.A.; Baduweh, C.A.; Sun, C.; Ding, Y. Functional Diagnostic and Therapeutic Nanoconstructs for Efficient Probing of Circulating Tumor Cells. *ACS Appl. Mater. Interfaces* **2018**, *10*, 14231–14247. [CrossRef]
45. Raimondi, C.; Gradilone, A.; Naso, G.; Cortesi, E.; Gazzaniga, P. Clinical utility of circulating tumor cell counting through CellSearch((R)): The dilemma of a concept suspended in Limbo. *Onco. Targets Ther.* **2014**, *7*, 619–625. [CrossRef]
46. Zhang, J.; Chen, K.; Fan, Z.H. Circulating Tumor Cell Isolation and Analysis. *Adv. Clin. Chem.* **2016**, *75*, 1–31. [CrossRef]
47. Alix-Panabieres, C.; Pantel, K. Technologies for detection of circulating tumor cells: Facts and vision. *Lab Chip* **2014**, *14*, 57–62. [CrossRef]
48. Wang, H.; Lin, Y.; Nienhaus, K.; Nienhaus, G.U. The protein corona on nanoparticles as viewed from a nanoparticle-sizing perspective. *Wiley Interdiscip. Rev. Nanomed. Nanobiotechnol.* **2017**. [CrossRef]
49. Monopoli, M.P.; Aberg, C.; Salvati, A.; Dawson, K.A. Biomolecular coronas provide the biological identity of nanosized materials. *Nat. Nanotechnol.* **2012**, *7*, 779–786. [CrossRef]
50. Monopoli, M.P.; Bombelli, F.B.; Dawson, K.A. Nanobiotechnology: Nanoparticle coronas take shape. *Nat. Nanotechnol.* **2011**, *6*, 11–12. [CrossRef]
51. Tenzer, S.; Docter, D.; Kuharev, J.; Musyanovych, A.; Fetz, V.; Hecht, R.; Schlenk, F.; Fischer, D.; Kiouptsi, K.; Reinhardt, C.; et al. Rapid formation of plasma protein corona critically affects nanoparticle pathophysiology. *Nat. Nanotechnol.* **2013**, *8*, 772–781. [CrossRef] [PubMed]
52. Treuel, L.; Docter, D.; Maskos, M.; Stauber, R.H. Protein corona—From molecular adsorption to physiological complexity. *Beilstein J. Nanotechnol.* **2015**, *6*, 857–873. [CrossRef] [PubMed]
53. Vroman, L. Effect of absorbed proteins on the wettability of hydrophilic and hydrophobic solids. *Nature* **1962**, *196*, 476–477. [CrossRef] [PubMed]
54. Cedervall, T.; Lynch, I.; Lindman, S.; Berggard, T.; Thulin, E.; Nilsson, H.; Dawson, K.A.; Linse, S. Understanding the nanoparticle-protein corona using methods to quantify exchange rates and affinities of proteins for nanoparticles. *Proc. Natl. Acad. Sci. USA* **2007**, *104*, 2050–2055. [CrossRef] [PubMed]
55. Westmeier, D.; Stauber, R.H.; Docter, D. The concept of bio-corona in modulating the toxicity of engineered nanomaterials (ENM). *Toxicol. Appl. Pharmacol.* **2016**, *299*, 53–57. [CrossRef] [PubMed]

56. Walczyk, D.; Bombelli, F.B.; Monopoli, M.P.; Lynch, I.; Dawson, K.A. What the cell "sees" in bionanoscience. *J. Am. Chem. Soc.* **2010**, *132*, 5761–5768. [CrossRef]
57. Walkey, C.D.; Olsen, J.B.; Song, F.; Liu, R.; Guo, H.; Olsen, D.W.; Cohen, Y.; Emili, A.; Chan, W.C. Protein corona fingerprinting predicts the cellular interaction of gold and silver nanoparticles. *ACS Nano* **2014**, *8*, 2439–2455. [CrossRef]
58. Myung, J.H.; Tam, K.A.; Park, S.J.; Cha, A.; Hong, S. Recent advances in nanotechnology-based detection and separation of circulating tumor cells. *Wiley Interdiscip. Rev. Nanomed. Nanobiotechnol.* **2016**, *8*, 223–239. [CrossRef]
59. Huang, Q.; Wang, Y.; Chen, X.; Wang, Y.; Li, Z.; Du, S.; Wang, L.; Chen, S. Nanotechnology-Based Strategies for Early Cancer Diagnosis Using Circulating Tumor Cells as a Liquid Biopsy. *Nanotheranostics* **2018**, *2*, 21–41. [CrossRef]
60. Mahmoudi, M.; Sant, S.; Wang, B.; Laurent, S.; Sen, T. Superparamagnetic iron oxide nanoparticles (SPIONs): Development, surface modification and applications in chemotherapy. *Adv. Drug Deliv. Rev.* **2011**, *63*, 24–46. [CrossRef]
61. Riethdorf, S.; O'Flaherty, L.; Hille, C.; Pantel, K. Clinical applications of the CellSearch platform in cancer patients. *Adv. Drug Deliv. Rev.* **2018**. [CrossRef] [PubMed]
62. Pamme, N. Magnetism and microfluidics. *Lab Chip* **2006**, *6*, 24–38. [CrossRef] [PubMed]
63. Rao, C.G.; Chianese, D.; Doyle, G.V.; Miller, M.C.; Russell, T.; Sanders, R.A., Jr.; Terstappen, L.W. Expression of epithelial cell adhesion molecule in carcinoma cells present in blood and primary and metastatic tumors. *Int. J. Oncol.* **2005**, *27*, 49–57. [CrossRef] [PubMed]
64. Torsten Schüling, A.E.; Thomas Scheperand and Johanna Walter. Aptamer-based lateral flow assays. *AIMS Bioeng.* **2018**, *5*, 78–102. [CrossRef]
65. Jia, Z.; Liang, Y.; Xu, X.; Li, X.; Liu, Q.; Ou, Y.; Duan, L.; Zhu, W.; Lu, W.; Xiong, J.; et al. Isolation and characterization of human mesenchymal stem cells derived from synovial fluid by magnetic-activated cell sorting (MACS). *Cell Biol. Int.* **2018**, *42*, 262–271. [CrossRef]
66. Andreopoulou, E.; Yang, L.Y.; Rangel, K.M.; Reuben, J.M.; Hsu, L.; Krishnamurthy, S.; Valero, V.; Fritsche, H.A.; Cristofanilli, M. Comparison of assay methods for detection of circulating tumor cells in metastatic breast cancer: AdnaGen AdnaTest BreastCancer Select/Detect versus Veridex CellSearch system. *Int. J. Cancer* **2012**, *130*, 1590–1597. [CrossRef]
67. Gorges, T.M.; Tinhofer, I.; Drosch, M.; Rose, L.; Zollner, T.M.; Krahn, T.; von Ahsen, O. Circulating tumour cells escape from EpCAM-based detection due to epithelial-to-mesenchymal transition. *BMC Cancer* **2012**, *12*, 178. [CrossRef]
68. Rosorius, O.; Heger, P.; Stelz, G.; Hirschmann, N.; Hauber, J.; Stauber, R.H. Direct observation of nucleocytoplasmic transport by microinjection of GFP-tagged proteins in living cells. *Biotechniques* **1999**, *27*, 350. [CrossRef]
69. Yu, M.; Stott, S.; Toner, M.; Maheswaran, S.; Haber, D.A. Circulating tumor cells: Approaches to isolation and characterization. *J. Cell Biol.* **2011**, *192*, 373–382. [CrossRef]
70. Lee, J.; Kang, H.J.; Jang, H.; Lee, Y.J.; Lee, Y.S.; Ali, B.A.; Al-Khedhairy, A.A.; Kim, S. Simultaneous imaging of two different cancer biomarkers using aptamer-conjugated quantum dots. *Sensors* **2015**, *15*, 8595–8604. [CrossRef]
71. Galanzha, E.I.; Zharov, V.P. Circulating Tumor Cell Detection and Capture by Photoacoustic Flow Cytometry in Vivo and ex Vivo. *Cancers* **2013**, *5*, 1691–1738. [CrossRef] [PubMed]
72. Wu, X.; Xia, Y.; Huang, Y.; Li, J.; Ruan, H.; Chen, T.; Luo, L.; Shen, Z.; Wu, A. Improved SERS-Active Nanoparticles with Various Shapes for CTC Detection without Enrichment Process with Supersensitivity and High Specificity. *ACS Appl. Mater. Interfaces* **2016**, *8*, 19928–19938. [CrossRef] [PubMed]
73. Cai, W.; Gao, T.; Hong, H.; Sun, J. Applications of gold nanoparticles in cancer nanotechnology. *Nanotechnol. Sci. Appl.* **2008**, *1*, 17–32. [CrossRef]
74. Hu, M.; Chen, J.; Li, Z.Y.; Au, L.; Hartland, G.V.; Li, X.; Marquez, M.; Xia, Y. Gold nanostructures: Engineering their plasmonic properties for biomedical applications. *Chem. Soc. Rev.* **2006**, *35*, 1084–1094. [CrossRef]
75. He, W.; Wang, H.; Hartmann, L.C.; Cheng, J.X.; Low, P.S. In vivo quantitation of rare circulating tumor cells by multiphoton intravital flow cytometry. *Proc. Natl. Acad. Sci. USA* **2007**, *104*, 11760–11765. [CrossRef] [PubMed]

76. He, B.; Yang, D.; Qin, M.; Zhang, Y.; He, B.; Dai, W.; Wang, X.; Zhang, Q.; Zhang, H.; Yin, C. Increased cellular uptake of peptide-modified PEGylated gold nanoparticles. *Biochem. Biophys. Res. Commun.* **2017**, *494*, 339–345. [CrossRef] [PubMed]
77. Zhang, Y.; Kohler, N.; Zhang, M. Surface modification of superparamagnetic magnetite nanoparticles and their intracellular uptake. *Biomaterials* **2002**, *23*, 1553–1561. [CrossRef]
78. Park, M.H.; Reategui, E.; Li, W.; Tessier, S.N.; Wong, K.H.; Jensen, A.E.; Thapar, V.; Ting, D.; Toner, M.; Stott, S.L.; et al. Enhanced Isolation and Release of Circulating Tumor Cells Using Nanoparticle Binding and Ligand Exchange in a Microfluidic Chip. *J. Am. Chem. Soc.* **2017**, *139*, 2741–2749. [CrossRef]
79. Pramani, K.A.; Jones, S.; Gao, Y.; Sweet, C.; Vangara, A.; Begum, S.; Ray, P.C. Multifunctional hybrid graphene oxide for circulating tumor cell isolation and analysis. *Adv. Drug Deliv. Rev.* **2018**. [CrossRef]
80. Yoon, H.J.; Kozminsky, M.; Nagrath, S. Emerging role of nanomaterials in circulating tumor cell isolation and analysis. *ACS Nano* **2014**, *8*, 1995–2017. [CrossRef]
81. Dreyer, D.R.; Park, S.; Bielawski, C.W.; Ruoff, R.S. The chemistry of graphene oxide. *Chem. Soc. Rev.* **2010**, *39*, 228–240. [CrossRef] [PubMed]
82. Sun, X.; Liu, Z.; Welsher, K.; Robinson, J.T.; Goodwin, A.; Zaric, S.; Dai, H. Nano-Graphene Oxide for Cellular Imaging and Drug Delivery. *Nano Res.* **2008**, *1*, 203–212. [CrossRef] [PubMed]
83. Yoon, H.J.; Kim, T.H.; Zhang, Z.; Azizi, E.; Pham, T.M.; Paoletti, C.; Lin, J.; Ramnath, N.; Wicha, M.S.; Hayes, D.F.; et al. Sensitive capture of circulating tumour cells by functionalized graphene oxide nanosheets. *Nat. Nanotechnol.* **2013**, *8*, 735–741. [CrossRef] [PubMed]
84. Loh, K.P.; Bao, Q.; Eda, G.; Chhowalla, M. Graphene oxide as a chemically tunable platform for optical applications. *Nat. Chem.* **2010**, *2*, 1015–1024. [CrossRef]
85. Wei, Z.; Barlow, D.E.; Sheehan, P.E. The assembly of single-layer graphene oxide and graphene using molecular templates. *Nano Lett.* **2008**, *8*, 3141–3145. [CrossRef]
86. Wu, Y.; Xue, P.; Kang, Y.; Hui, K.M. Highly specific and ultrasensitive graphene-enhanced electrochemical detection of low-abundance tumor cells using silica nanoparticles coated with antibody-conjugated quantum dots. *Anal. Chem.* **2013**, *85*, 3166–3173. [CrossRef]
87. Murray, A.R.; Kisin, E.R.; Tkach, A.V.; Yanamala, N.; Mercer, R.; Young, S.H.; Fadeel, B.; Kagan, V.E.; Shvedova, A.A. Factoring-in agglomeration of carbon nanotubes and nanofibers for better prediction of their toxicity versus asbestos. *Part. Fibre Toxicol.* **2012**, *9*, 10. [CrossRef]
88. Allegri, M.; Perivoliotis, D.K.; Bianchi, M.G.; Chiu, M.; Pagliaro, A.; Koklioti, M.A.; Trompeta, A.A.; Bergamaschi, E.; Bussolati, O.; Charitidis, C.A. Toxicity determinants of multi-walled carbon nanotubes: The relationship between functionalization and agglomeration. *Toxicol. Rep.* **2016**, *3*, 230–243. [CrossRef]
89. De Jong, K.P.; Geus, J.W. Carbon nanofibers: Catalytic synthesis and applications. *Catal. Rev. Sci. Eng.* **2000**, *42*, 481–510. [CrossRef]
90. Shao, N.; Wickstrom, E.; Panchapakesan, B. Nanotube-antibody biosensor arrays for the detection of circulating breast cancer cells. *Nanotechnology* **2008**, *19*, 465101. [CrossRef]
91. Liu, Y.; Zhu, F.; Dan, W.; Fu, Y.; Liu, S. Construction of carbon nanotube based nanoarchitectures for selective impedimetric detection of cancer cells in whole blood. *Analyst* **2014**, *139*, 5086–5092. [CrossRef] [PubMed]
92. Tan, S.J.; Yobas, L.; Lee, G.Y.; Ong, C.N.; Lim, C.T. Microdevice for the isolation and enumeration of cancer cells from blood. *Biomed. Microdevices* **2009**, *11*, 883–892. [CrossRef] [PubMed]
93. Zheng, S.; Lin, H.; Liu, J.Q.; Balic, M.; Datar, R.; Cote, R.J.; Tai, Y.C. Membrane microfilter device for selective capture, electrolysis and genomic analysis of human circulating tumor cells. *J. Chromatogr. A* **2007**, *1162*, 154–161. [CrossRef] [PubMed]
94. Harvey, C. Ultrasound with microbubbles. *Cancer Imaging* **2015**, *15*, O19. [CrossRef]
95. Ambika Rajendran, M. Ultrasound-guided Microbubble in the Treatment of Cancer: A Mini Narrative Review. *Cureus* **2018**, *10*, e3256. [CrossRef]
96. Schutt, E.G.; Klein, D.H.; Mattrey, R.M.; Riess, J.G. Injectable microbubbles as contrast agents for diagnostic ultrasound imaging: The key role of perfluorochemicals. *Angew. Chem. Int. Ed. Engl.* **2003**, *42*, 3218–3235. [CrossRef]
97. Talu, E.; Hettiarachchi, K.; Powell, R.L.; Lee, A.P.; Dayton, P.A.; Longo, M.L. Maintaining monodispersity in a microbubble population formed by flow-focusing. *Langmuir* **2008**, *24*, 1745–1749. [CrossRef]
98. Sirsi, S.; Borden, M. Microbubble Compositions, Properties and Biomedical Applications. *Bubble Sci. Eng. Technol.* **2009**, *1*, 3–17. [CrossRef]

99. Szijjarto, C.; Rossi, S.; Waton, G.; Krafft, M.P. Effects of perfluorocarbon gases on the size and stability characteristics of phospholipid-coated microbubbles: Osmotic effect versus interfacial film stabilization. *Langmuir* **2012**, *28*, 1182–1189. [CrossRef]
100. Yeh, J.S.; Sennoga, C.A.; McConnell, E.; Eckersley, R.; Tang, M.X.; Nourshargh, S.; Seddon, J.M.; Haskard, D.O.; Nihoyannopoulos, P. A Targeting Microbubble for Ultrasound Molecular Imaging. *PLoS ONE* **2015**, *10*, e0129681. [CrossRef]
101. Wang, S.; Hossack, J.A.; Klibanov, A.L. Targeting of microbubbles: Contrast agents for ultrasound molecular imaging. *J. Drug Target* **2018**, *26*, 420–434. [CrossRef] [PubMed]
102. Wang, G.; Benasutti, H.; Jones, J.F.; Shi, G.; Benchimol, M.; Pingle, S.; Kesari, S.; Yeh, Y.; Hsieh, L.E.; Liu, Y.T.; et al. Isolation of Breast cancer CTCs with multitargeted buoyant immunomicrobubbles. *Colloids Surf. B Biointerfaces* **2018**, *161*, 200–209. [CrossRef] [PubMed]
103. Abou-Elkacem, L.; Bachawal, S.V.; Willmann, J.K. Ultrasound molecular imaging: Moving toward clinical translation. *Eur. J. Radiol.* **2015**, *84*, 1685–1693. [CrossRef] [PubMed]
104. Smeenge, M.; Tranquart, F.; Mannaerts, C.K.; de Reijke, T.M.; van de Vijver, M.J.; Laguna, M.P.; Pochon, S.; de la Rosette, J.; Wijkstra, H. First-in-Human Ultrasound Molecular Imaging With a VEGFR2-Specific Ultrasound Molecular Contrast Agent (BR55) in Prostate Cancer: A Safety and Feasibility Pilot Study. *Investig. Radiol.* **2017**, *52*, 419–427. [CrossRef]
105. Willmann, J.K.; Bonomo, L.; Testa, A.C.; Rinaldi, P.; Rindi, G.; Valluru, K.S.; Petrone, G.; Martini, M.; Lutz, A.M.; Gambhir, S.S. Ultrasound Molecular Imaging With BR55 in Patients With Breast and Ovarian Lesions: First-in-Human Results. *J. Clin. Oncol.* **2017**, *35*, 2133–2140. [CrossRef]
106. Chong, W.K.; Papadopoulou, V.; Dayton, P.A. Imaging with ultrasound contrast agents: Current status and future. *Abdom. Radiol. (NY)* **2018**, *43*, 762–772. [CrossRef]
107. Ferrara, K.W.; Borden, M.A.; Zhang, H. Lipid-shelled vehicles: Engineering for ultrasound molecular imaging and drug delivery. *Acc. Chem. Res.* **2009**, *42*, 881–892. [CrossRef]
108. Shi, G.; Cui, W.; Mukthavaram, R.; Liu, Y.T.; Simberg, D. Binding and isolation of tumor cells in biological media with perfluorocarbon microbubbles. *Methods* **2013**, *64*, 102–107. [CrossRef]
109. Shi, G.X.; Cui, W.J.; Benchimol, M.; Liu, Y.T.; Mattrey, R.F.; Mukthavaram, R.; Kesari, S.; Esener, S.C.; Simberg, D. Isolation of Rare Tumor Cells from Blood Cells with Buoyant Immuno-Microbubbles. *PLoS ONE* **2013**, *8*. [CrossRef]
110. Hernot, S.; Klibanov, A.L. Microbubbles in ultrasound-triggered drug and gene delivery. *Adv. Drug Deliv. Rev.* **2008**, *60*, 1153–1166. [CrossRef]
111. Jain, A.; Tiwari, A.; Verma, A.; Jain, S.K. Ultrasound-based triggered drug delivery to tumors. *Drug Deliv. Transl. Res.* **2018**, *8*, 150–164. [CrossRef] [PubMed]
112. Unger, E.C.; Porter, T.; Culp, W.; Labell, R.; Matsunaga, T.; Zutshi, R. Therapeutic applications of lipid-coated microbubbles. *Adv. Drug Deliv. Rev.* **2004**, *56*, 1291–1314. [CrossRef] [PubMed]
113. Aw, M.S.; Paniwnyk, L.; Losic, D. The progressive role of acoustic cavitation for non-invasive therapies, contrast imaging and blood-tumor permeability enhancement. *Expert Opin. Drug Deliv.* **2016**, *13*, 1383–1396. [CrossRef] [PubMed]
114. Song, K.H.; Harvey, B.K.; Borden, M.A. State-of-the-art of microbubble-assisted blood-brain barrier disruption. *Theranostics* **2018**, *8*, 4393–4408. [CrossRef] [PubMed]
115. Simberg, D.; Mattrey, R. Targeting of perfluorocarbon microbubbles to selective populations of circulating blood cells. *J. Drug Target* **2009**, *17*, 392–398. [CrossRef] [PubMed]
116. Hsu, C.H.; Chen, C.; Irimia, D.; Toner, M. Isolating cells from blood using buoyancy activated cell sorting (BACS) with glass microbubbles. In Proceedings of the 14th International Conference on Miniaturized Systems for Chemistry and Life Sciences, Groningen, The Netherlands, 3–7 October 2010.
117. Cristofanilli, M.; Budd, G.T.; Ellis, M.J.; Stopeck, A.; Matera, J.; Miller, M.C.; Reuben, J.M.; Doyle, G.V.; Allard, W.J.; Terstappen, L.W.M.M.; et al. Circulating tumor cells, disease progression, and survival in metastatic breast cancer. *New Engl. J. Med.* **2004**, *351*, 781–791. [CrossRef]
118. Liou, Y.R.; Wang, Y.H.; Lee, C.Y.; Li, P.C. Buoyancy-activated cell sorting using targeted biotinylated albumin microbubbles. *PLoS ONE* **2015**, *10*, e0125036. [CrossRef]
119. Zhu, L.; Cheng, G.; Ye, D.; Nazeri, A.; Yue, Y.; Liu, W.; Wang, X.; Dunn, G.P.; Petti, A.A.; Leuthardt, E.C.; et al. Focused Ultrasound-enabled Brain Tumor Liquid Biopsy. *Sci. Rep.* **2018**, *8*, 6553. [CrossRef]

120. Schmitz, B.; Radbruch, A.; Kummel, T.; Wickenhauser, C.; Korb, H.; Hansmann, M.L.; Thiele, J.; Fischer, R. Magnetic activated cell sorting (MACS)—A new immunomagnetic method for megakaryocytic cell isolation: Comparison of different separation techniques. *Eur. J. Haematol.* **1994**, *52*, 267–275. [CrossRef]
121. Macey, M.G. Flow cytometry: Principles and clinical applications. *Med. Lab Sci.* **1988**, *45*, 165–173. [PubMed]
122. Bianchi, D.W.; Klinger, K.W.; Vadnais, T.J.; Demaria, M.A.; Shuber, A.P.; Skoletsky, J.; Midura, P.; Diriso, M.; Pelletier, C.; Genova, M.; et al. Development of a model system to compare cell separation methods for the isolation of fetal cells from maternal blood. *Prenat. Diagn.* **1996**, *16*, 289–298. [CrossRef]
123. Vankooten, T.G.; Schakenraad, J.M.; Vandermei, H.C.; Dekker, A.; Kirkpatrick, C.J.; Busscher, H.J. Fluid Shear-Induced Endothelial-Cell Detachment from Glass—Influence of Adhesion Time and Shear-Stress. *Med. Eng. Phys.* **1994**, *16*, 506–512. [CrossRef]
124. Wu, J. Mechanisms of animal cell damage associated with gas bubbles and cell protection by medium additives. *J. Biotechnol.* **1995**, *43*, 81–94. [CrossRef]
125. Krivacic, R.T.; Ladanyi, A.; Curry, D.N.; Hsieh, H.B.; Kuhn, P.; Bergsrud, D.E.; Kepros, J.F.; Barbera, T.; Ho, M.Y.; Chen, L.B.; et al. A rare-cell detector for cancer. *Proc. Natl. Acad. Sci. USA* **2004**, *101*, 10501–10504. [CrossRef] [PubMed]
126. Kraeft, S.K.; Ladanyi, A.; Galiger, K.; Herlitz, A.; Sher, A.C.; Bergsrud, D.E.; Even, G.; Brunelle, S.; Harris, L.; Salgia, R.; et al. Reliable and sensitive identification of occult tumor cells using the improved rare event imaging system. *Clin. Cancer Res.* **2004**, *10*, 3020–3028. [CrossRef]
127. Bauer, K.D.; de la Torre-Bueno, J.; Diel, I.J.; Hawes, D.; Decker, W.J.; Priddy, C.; Bossy, B.; Ludmann, S.; Yamamoto, K.; Masih, A.S.; et al. Reliable and sensitive analysis of occult bone marrow metastases using automated cellular imaging. *Clin. Cancer Res.* **2000**, *6*, 3552–3559.
128. Owen, J.; Crake, C.; Lee, J.Y.; Carugo, D.; Beguin, E.; Khrapitchev, A.A.; Browning, R.J.; Sibson, N.; Stride, E. A versatile method for the preparation of particle-loaded microbubbles for multimodality imaging and targeted drug delivery. *Drug Deliv. Transl. Res.* **2018**, *8*, 342–356. [CrossRef]
129. Park, J.I.; Jagadeesan, D.; Williams, R.; Oakden, W.; Chung, S.; Stanisz, G.J.; Kumacheva, E. Microbubbles loaded with nanoparticles: A route to multiple imaging modalities. *ACS Nano* **2010**, *4*, 6579–6586. [CrossRef]
130. Jin, B.; Lin, M.; Zong, Y.; Wan, M.; Xu, F.; Duan, Z.; Lu, T. Microbubble embedded with upconversion nanoparticles as a bimodal contrast agent for fluorescence and ultrasound imaging. *Nanotechnology* **2015**, *26*, 345601. [CrossRef]
131. Dong, Z.; Yu, D.; Liu, Q.; Ding, Z.; Lyons, V.J.; Bright, R.K.; Pappas, D.; Liu, X.; Li, W. Enhanced capture and release of circulating tumor cells using hollow glass microspheres with a nanostructured surface. *Nanoscale* **2018**, *10*, 16795–16804. [CrossRef] [PubMed]
132. Wei, Y.; Liao, R.; Mahmood, A.A.; Xu, H.; Zhou, Q. pH-responsive pHLIP (pH low insertion peptide) nanoclusters of superparamagnetic iron oxide nanoparticles as a tumor-selective MRI contrast agent. *Acta Biomater.* **2017**, *55*, 194–203. [CrossRef] [PubMed]
133. Jalani, G.; Jeyachandran, D.; Bertram Church, R.; Cerruti, M. Graphene oxide-stabilized perfluorocarbon emulsions for controlled oxygen delivery. *Nanoscale* **2017**, *9*, 10161–10166. [CrossRef] [PubMed]
134. Justeau, C.; Vela-Gonzalez, A.V.; Jourdan, A.; Riess, J.G.; Krafft, M.P. Adsorption of Cerium Salts and Cerium Oxide Nanoparticles on Microbubbles Can Be Induced by a Fluorocarbon Gas. *ACS Sustain. Chem. Eng.* **2018**, *6*, 11450–11456. [CrossRef]
135. McLaughlan, J.R.; Harput, S.; Abou-Saleh, R.H.; Peyman, S.A.; Evans, S.; Freear, S. Characterisation of Liposome-Loaded Microbubble Populations for Subharmonic Imaging. *Ultrasound Med. Biol.* **2017**, *43*, 346–356. [CrossRef]
136. Shi, D.; Wallyn, J.; Nguyen, D.-V.; Perton, F.; Felder-Flesch, D.; Bégin-Colin, S.; Maaloum, M.; Krafft, M.P. Microbubbles decorated with dendronized magnetic nanoparticles for biomedical imaging. Effective stabilization via fluorous interactions. *Beilstein J. Nanotechnol.* **2019**, *10*, 2103–2115. [CrossRef]
137. Qian, W.; Zhang, Y.; Chen, W. Capturing Cancer: Emerging Microfluidic Technologies for the Capture and Characterization of Circulating Tumor Cells. *Small* **2015**, *11*, 3850–3872. [CrossRef]
138. Jackson, J.M.; Witek, M.A.; Kamande, J.W.; Soper, S.A. Materials and microfluidics: Enabling the efficient isolation and analysis of circulating tumour cells. *Chem. Soc. Rev.* **2017**, *46*, 4245–4280. [CrossRef]
139. Li, P.; Stratton, Z.S.; Dao, M.; Ritz, J.; Huang, T.J. Probing circulating tumor cells in microfluidics. *Lab Chip* **2013**, *13*, 602–609. [CrossRef]

140. Hoshino, K.; Huang, Y.Y.; Lane, N.; Huebschman, M.; Uhr, J.W.; Frenkel, E.P.; Zhang, X. Microchip-based immunomagnetic detection of circulating tumor cells. *Lab Chip* **2011**, *11*, 3449–3457. [CrossRef]
141. Huang, Y.Y.; Hoshino, K.; Chen, P.; Wu, C.H.; Lane, N.; Huebschman, M.; Liu, H.; Sokolov, K.; Uhr, J.W.; Frenkel, E.P.; et al. Immunomagnetic nanoscreening of circulating tumor cells with a motion controlled microfluidic system. *Biomed. Microdevices* **2013**, *15*, 673–681. [CrossRef] [PubMed]
142. Issadore, D.; Chung, J.; Shao, H.; Liong, M.; Ghazani, A.A.; Castro, C.M.; Weissleder, R.; Lee, H. Ultrasensitive clinical enumeration of rare cells ex vivo using a micro-hall detector. *Sci. Transl. Med.* **2012**, *4*, 141–192. [CrossRef] [PubMed]
143. Nagrath, S.; Sequist, L.V.; Maheswaran, S.; Bell, D.W.; Irimia, D.; Ulkus, L.; Smith, M.R.; Kwak, E.L.; Digumarthy, S.; Muzikansky, A.; et al. Isolation of rare circulating tumour cells in cancer patients by microchip technology. *Nature* **2007**, *450*, 1235–1239. [CrossRef] [PubMed]
144. Stott, S.L.; Hsu, C.H.; Tsukrov, D.I.; Yu, M.; Miyamoto, D.T.; Waltman, B.A.; Rothenberg, S.M.; Shah, A.M.; Smas, M.E.; Korir, G.K.; et al. Isolation of circulating tumor cells using a microvortex-generating herringbone-chip. *Proc. Natl. Acad. Sci. USA* **2010**, *107*, 18392–18397. [CrossRef]
145. McCarley, R.L.; Vaidya, B.; Wei, S.; Smith, A.F.; Patel, A.B.; Feng, J.; Murphy, M.C.; Soper, S.A. Resist-free patterning of surface architectures in polymer-based microanalytical devices. *J. Am. Chem. Soc.* **2005**, *127*, 842–843. [CrossRef]
146. Adams, A.A.; Okagbare, P.I.; Feng, J.; Hupert, M.L.; Patterson, D.; Gottert, J.; McCarley, R.L.; Nikitopoulos, D.; Murphy, M.C.; Soper, S.A. Highly efficient circulating tumor cell isolation from whole blood and label-free enumeration using polymer-based microfluidics with an integrated conductivity sensor. *J. Am. Chem. Soc.* **2008**, *130*, 8633–8641. [CrossRef]
147. Wang, Y.; Zhao, Q.; Han, N.; Bai, L.; Li, J.; Liu, J.; Che, E.; Hu, L.; Zhang, Q.; Jiang, T.; et al. Mesoporous silica nanoparticles in drug delivery and biomedical applications. *Nanomedicine* **2015**, *11*, 313–327. [CrossRef]
148. Gao, Y.; Gu, S.; Zhang, Y.; Xie, X.; Yu, T.; Lu, Y.; Zhu, Y.; Chen, W.; Zhang, H.; Dong, H.; et al. The Architecture and Function of Monoclonal Antibody-Functionalized Mesoporous Silica Nanoparticles Loaded with Mifepristone: Repurposing Abortifacient for Cancer Metastatic Chemoprevention. *Small* **2016**, *12*, 2595–2608. [CrossRef]
149. Mahapatro, A.; Singh, D.K. Biodegradable nanoparticles are excellent vehicle for site directed in-vivo delivery of drugs and vaccines. *J. Nanobiotechnology* **2011**, *9*, 55. [CrossRef]
150. Peppas, N.A. Historical perspective on advanced drug delivery: How engineering design and mathematical modeling helped the field mature. *Adv. Drug Deliv. Rev.* **2013**, *65*, 5–9. [CrossRef]
151. Croy, S.R.; Kwon, G.S. Polymeric micelles for drug delivery. *Curr. Pharm. Des.* **2006**, *12*, 4669–4684. [CrossRef] [PubMed]
152. Jeong, B.; Bae, Y.H.; Lee, D.S.; Kim, S.W. Biodegradable block copolymers as injectable drug-delivery systems. *Nature* **1997**, *388*, 860–862. [CrossRef] [PubMed]
153. Patri, A.K.; Majoros, I.J.; Baker, J.R. Dendritic polymer macromolecular carriers for drug delivery. *Curr. Opin. Chem. Biol.* **2002**, *6*, 466–471. [CrossRef]
154. Gaucher, G.; Dufresne, M.H.; Sant, V.P.; Kang, N.; Maysinger, D.; Leroux, J.C. Block copolymer micelles: Preparation, characterization and application in drug delivery. *J. Control Release* **2005**, *109*, 169–188. [CrossRef] [PubMed]
155. Langer, R.; Folkman, J. Polymers for the sustained release of proteins and other macromolecules. *Nature* **1976**, *263*, 797–800. [CrossRef] [PubMed]
156. Sadat Tabatabaei Mirakabad, F.; Nejati-Koshki, K.; Akbarzadeh, A.; Yamchi, M.R.; Milani, M.; Zarghami, N.; Zeighamian, V.; Rahimzadeh, A.; Alimohammadi, S.; Hanifehpour, Y.; et al. PLGA-based nanoparticles as cancer drug delivery systems. *Asian Pac. J. Cancer Prev.* **2014**, *15*, 517–535. [CrossRef] [PubMed]
157. Jain, R.A. The manufacturing techniques of various drug loaded biodegradable poly(lactide-co-glycolide) (PLGA) devices. *Biomaterials* **2000**, *21*, 2475–2490. [CrossRef]
158. Lu, J.M.; Wang, X.; Marin-Muller, C.; Wang, H.; Lin, P.H.; Yao, Q.; Chen, C. Current advances in research and clinical applications of PLGA-based nanotechnology. *Expert Rev. Mol. Diagn.* **2009**, *9*, 325–341. [CrossRef]
159. Studer, M.; Briel, M.; Leimenstoll, B.; Glass, T.R.; Bucher, H.C. Effect of different antilipidemic agents and diets on mortality: A systematic review. *Arch. Intern. Med.* **2005**, *165*, 725–730. [CrossRef]
160. Wickline, S.A.; Neubauer, A.M.; Winter, P.M.; Caruthers, S.D.; Lanza, G.M. Molecular imaging and therapy of atherosclerosis with targeted nanoparticles. *J. Magn. Reson. Imaging* **2007**, *25*, 667–680. [CrossRef]

161. Trivedi, R.; Kompella, U.B. Nanomicellar formulations for sustained drug delivery: Strategies and underlying principles. *Nanomedicine (Lond)* **2010**, *5*, 485–505. [CrossRef] [PubMed]
162. Venkatraman, S.S.; Jie, P.; Min, F.; Freddy, B.Y.; Leong-Huat, G. Micelle-like nanoparticles of PLA-PEG-PLA triblock copolymer as chemotherapeutic carrier. *Int. J. Pharm.* **2005**, *298*, 219–232. [CrossRef] [PubMed]
163. Bisht, S.; Maitra, A. Dextran-doxorubicin/chitosan nanoparticles for solid tumor therapy. *Wiley Interdiscip. Rev. Nanomed. Nanobiotechnol.* **2009**, *1*, 415–425. [CrossRef] [PubMed]
164. Ye, Y.Q.; Yang, F.L.; Hu, F.Q.; Du, Y.Z.; Yuan, H.; Yu, H.Y. Core-modified chitosan-based polymeric micelles for controlled release of doxorubicin. *Int. J. Pharm.* **2008**, *352*, 294–301. [CrossRef]
165. Deng, S.; Wu, Q.; Zhao, Y.; Zheng, X.; Wu, N.; Pang, J.; Li, X.; Bi, C.; Liu, X.; Yang, L.; et al. Biodegradable polymeric micelle-encapsulated doxorubicin suppresses tumor metastasis by killing circulating tumor cells. *Nanoscale* **2015**, *7*, 5270–5280. [CrossRef]
166. Yao, J.; Feng, J.; Gao, X.; Wei, D.; Kang, T.; Zhu, Q.; Jiang, T.; Wei, X.; Chen, J. Neovasculature and circulating tumor cells dual-targeting nanoparticles for the treatment of the highly-invasive breast cancer. *Biomaterials* **2017**, *113*, 1–17. [CrossRef]
167. Gener, P.; Montero, S.; Xandri-Monje, H.; Diaz-Riascos, Z.V.; Rafael, D.; Andrade, F.; Martinez-Trucharte, F.; Gonzalez, P.; Seras-Franzoso, J.; Manzano, A.; et al. Zileuton loaded in polymer micelles effectively reduce breast cancer circulating tumor cells and intratumoral cancer stem cells. *Nanomedicine* **2019**, *24*, 102106. [CrossRef]
168. Medina, S.H.; El-Sayed, M.E. Dendrimers as carriers for delivery of chemotherapeutic agents. *Chem. Rev.* **2009**, *109*, 3141–3157. [CrossRef]
169. Xie, J.; Dong, H.; Chen, H.; Zhao, R.; Sinko, P.J.; Shen, W.; Wang, J.; Lu, Y.; Yang, X.; Xie, F.; et al. Exploring cancer metastasis prevention strategy: Interrupting adhesion of cancer cells to vascular endothelia of potential metastatic tissues by antibody-coated nanomaterial. *J. Nanobiotechnology* **2015**, *13*, 9. [CrossRef]
170. Xie, J.; Gao, Y.; Zhao, R.; Sinko, P.J.; Gu, S.; Wang, J.; Li, Y.; Lu, Y.; Yu, S.; Wang, L.; et al. Ex vivo and in vivo capture and deactivation of circulating tumor cells by dual-antibody-coated nanomaterials. *J. Control Release* **2015**, *209*, 159–169. [CrossRef]
171. Xie, J.; Zhao, R.; Gu, S.; Dong, H.; Wang, J.; Lu, Y.; Sinko, P.J.; Yu, T.; Xie, F.; Wang, L.; et al. The architecture and biological function of dual antibody-coated dendrimers: Enhanced control of circulating tumor cells and their hetero-adhesion to endothelial cells for metastasis prevention. *Theranostics* **2014**, *4*, 1250–1263. [CrossRef] [PubMed]
172. Brandl, A.; Wagner, T.; Uhlig, K.M.; Knauer, S.K.; Stauber, R.H.; Melchior, F.; Schneider, G.; Heinzel, T.; Kramer, O.H. Dynamically regulated sumoylation of HDAC2 controls p53 deacetylation and restricts apoptosis following genotoxic stress. *J. Mol. Cell Biol.* **2012**, *4*, 284–293. [CrossRef] [PubMed]
173. Dong, H.; Han, L.; Wu, Z.-S.; Zhang, T.; Xie, J.; Ma, J.; Wang, J.; Li, T.; Gao, Y.; Shao, J.; et al. Biostable Aptamer Rings Conjugated for Targeting Two Biomarkers on Circulating Tumor Cells in Vivo with Great Precision. *Chem. Mater.* **2017**, *24*, 10312–10325. [CrossRef]
174. Zheng, F.; Cheng, Y.; Wang, J.; Lu, J.; Zhang, B.; Zhao, Y.; Gu, Z. Aptamer-functionalized barcode particles for the capture and detection of multiple types of circulating tumor cells. *Adv. Mater.* **2014**, *26*, 7333–7338. [CrossRef]
175. Wood, A.K.; Sehgal, C.M. A review of low-intensity ultrasound for cancer therapy. *Ultrasound Med. Biol.* **2015**, *41*, 905–928. [CrossRef]
176. Heath, C.H.; Sorace, A.; Knowles, J.; Rosenthal, E.; Hoyt, K. Microbubble therapy enhances anti-tumor properties of cisplatin and cetuximab in vitro and in vivo. *Otolaryngol. Head Neck Surg.* **2012**, *146*, 938–945. [CrossRef]
177. Burgess, A.; Hynynen, K. Drug delivery across the blood-brain barrier using focused ultrasound. *Expert Opin. Drug. Deliv.* **2014**, *11*, 711–721. [CrossRef]
178. Mayer, C.R.; Geis, N.A.; Katus, H.A.; Bekeredjian, R. Ultrasound targeted microbubble destruction for drug and gene delivery. *Expert Opin. Drug Deliv.* **2008**, *5*, 1121–1138. [CrossRef]
179. Liu, Y.; Miyoshi, H.; Nakamura, M. Encapsulated ultrasound microbubbles: Therapeutic application in drug/gene delivery. *J. Control Release* **2006**, *114*, 89–99. [CrossRef]
180. Stride, E.P.; Coussios, C.C. Cavitation and contrast: The use of bubbles in ultrasound imaging and therapy. *Proc. Inst. Mech. Eng. H* **2010**, *224*, 171–191. [CrossRef]

181. Zhao, Y.Z.; Du, L.N.; Lu, C.T.; Jin, Y.G.; Ge, S.P. Potential and problems in ultrasound-responsive drug delivery systems. *Int. J. Nanomed.* **2013**, *8*, 1621–1633. [CrossRef] [PubMed]
182. Zheng, Y.; Zhang, Y.; Ao, M.; Zhang, P.; Zhang, H.; Li, P.; Qing, L.; Wang, Z.; Ran, H. Hematoporphyrin encapsulated PLGA microbubble for contrast enhanced ultrasound imaging and sonodynamic therapy. *J. Microencapsul.* **2012**, *29*, 437–444. [CrossRef] [PubMed]
183. Misik, V.; Riesz, P. Free radical intermediates in sonodynamic therapy. *Ann. N Y Acad. Sci.* **2000**, *899*, 335–348. [CrossRef]
184. Kuroki, M.; Hachimine, K.; Abe, H.; Shibaguchi, H.; Kuroki, M.; Maekawa, S.; Yanagisawa, J.; Kinugasa, T.; Tanaka, T.; Yamashita, Y. Sonodynamic therapy of cancer using novel sonosensitizers. *Anticancer Res.* **2007**, *27*, 3673–3677. [PubMed]
185. McEwan, C.; Fowley, C.; Nomikou, N.; McCaughan, B.; McHale, A.P.; Callan, J.F. Polymeric microbubbles as delivery vehicles for sensitizers in sonodynamic therapy. *Langmuir* **2014**, *30*, 14926–14930. [CrossRef]
186. McEwan, C.; Kamila, S.; Owen, J.; Nesbitt, H.; Callan, B.; Borden, M.; Nomikou, N.; Hamoudi, R.A.; Taylor, M.A.; Stride, E.; et al. Combined sonodynamic and antimetabolite therapy for the improved treatment of pancreatic cancer using oxygen loaded microbubbles as a delivery vehicle. *Biomaterials* **2016**, *80*, 20–32. [CrossRef]
187. McEwan, C.; Owen, J.; Stride, E.; Fowley, C.; Nesbitt, H.; Cochrane, D.; Coussios, C.C.; Borden, M.; Nomikou, N.; McHale, A.P.; et al. Oxygen carrying microbubbles for enhanced sonodynamic therapy of hypoxic tumours. *J. Control Release* **2015**, *203*, 51–56. [CrossRef]
188. Levenback, B.J.; Sehgal, C.M.; Wood, A.K. Modeling of thermal effects in antivascular ultrasound therapy. *J. Acoust. Soc. Am.* **2012**, *131*, 540–549. [CrossRef]
189. Geers, B.; De Wever, O.; Demeester, J.; Bracke, M.; De Smedt, S.C.; Lentacker, I. Targeted liposome-loaded microbubbles for cell-specific ultrasound-triggered drug delivery. *Small* **2013**, *9*, 4027–4035. [CrossRef]
190. Etheridge, M.L.; Campbell, S.A.; Erdman, A.G.; Haynes, C.L.; Wolf, S.M.; McCullough, J. The big picture on nanomedicine: The state of investigational and approved nanomedicine products. *Nanomedicine* **2013**, *9*, 1–14. [CrossRef]

MDPI
St. Alban-Anlage 66
4052 Basel
Switzerland
Tel. +41 61 683 77 34
Fax +41 61 302 89 18
www.mdpi.com

Nanomaterials Editorial Office
E-mail: nanomaterials@mdpi.com
www.mdpi.com/journal/nanomaterials

www.ingramcontent.com/pod-product-compliance
Lightning Source LLC
LaVergne TN
LVHW070705170726
843515LV00004B/494

* 9 7 8 3 0 3 6 5 2 7 3 9 0 *